THE LAST
VISITATION

JAMES MAURICE IBER

BLURBS

Retired Astrophysicist and U.S. Black Ops program scientist Dr. Horace Mitchel has been placed on an international government hit list. He's on the list because he's in possession of highly classified information, information that would literally change the world's entire traditionally accepted religious and historical views and destroy the system.

He possesses above top-secret information that reveals how we've always had "Visitations" from the negroid looking, black aliens who created us. In his old leather journal is secret information about how these interdimensional entities have assisted by giving us new technology and on occasion intervening in the affairs of men.

Dr. Mitchel has detailed evidence that proves that the 1947 Roswell New Mexico so-called alien crashes were faked, to give birth to the lie of "little green men" and flying saucers. His journal holds the proof that exposes how the governments of the world, led by the United States of America, have conspired to never allow the masses to know the truth. The truth is we are not alone and never have been.

Dr. Mitchel and the twelve people who eventually join him, go on an amazing journey to where mankind's history really begins. A distant place called Adimowa, which we call Heaven.

CONTENTS

"This book is dedicated to my father "Gene." My Dad passed away from COVID-19 complications at the age of 83, in Lancaster, California On September 7, 2021. He was laid to rest on September 27, 2021, at 2:00 PM. My Dad was a very special and talented published author, actor, singer, published songwriter, excellent painter, Olympiad, and much more. He truly inspired me to be all that I can be and to not only be a "Jack of all trades"; but preferably, a master of every single one. Dad, our love for you is eternal as the spirit of man. I will see you again." Revelations 22:14

"I must give a very special thanks to my wife and beautiful children for always being there and affording me the time it takes to write. My loves, thank you for the sacrifices you have made throughout the years."
James Maurice Iber

CHAPTER 1

THE JOURNEYS BEGIN

VALLEY HOSPITAL MEDICAL CENTER, LAS VEGAS NEVADA, JULY 9, 2021 7:00 PM (PT)

Everyone that has been screened and cleared to be on the COVID floor tenses up in anguish, as another *Code Blue* goes out over the PA system. A team of overworked and exhausted nurses run down the cold hallway to a man who fights for his life in the intensive care unit. The Doctor walks into the operating room and quickly looks at the man's chart. "We're losing him!" he shouts. He's going into cardiac arrest. Fight Mr. Baker, he yells! They try to jump-start his heart like an old car battery; they give him a shock from a defibrillator. "Clear, do it again!" They do, but it doesn't resuscitate him. His nearly lifeless body jumps off the table from the electricity that surges through it, yet no signs of life. The ominous tone of a flatline rings out. Carl Baker has given up the ghost, as they say. There's nothing more final than that sound.

"He's gone, turn that thing off," the Doctor says, holding his head in his hand. Another life lost to the pandemic and another mental scar for the doctor and nurses to carry for life. They all remove their gloves and scrubs. "Call it," the Doctor says. "Time of death, 7:11 PM," a nurse says. "Ok, let's wrap up here." The Doctor walks over to the door, then, turning toward everyone, "I'll go ruin another family's life." Some of the nurses drop their heads, others just walk away cold. "I hate this shit," one of them says. The ICU goes silent, empty, and dark. It looks like quite a battle has just taken place, with rubber gloves, tubes, and plastic wrappers all over the floor. On the ceiling, in the corner of the room, is a tiny speck of light hovering in the air. After a few moments, as if to say goodbye to his own body, it's time to go. In a flash, his journey to the next life begins.

United States Airforce Facility,

Area 51, (Sector 1)

Nevada "The Extraterrestrial Highway",

July 9, 2021, 7:11 PM (PT)

A thick cloud of fire and vapor appears and moves across the already sweltering Nevada desert, at exactly 7:11 in the evening. It approaches the restricted airspace of Area 51 and is closely monitored and tracked by a watchful Space Force Airman. The cloud stops directly over the base and hovers for hours. People from all ranks and levels leave their cubicles and offices just to see this strange event unfolding above them. The Secretary of the Airforce, Frank Matsuri, understands exactly what he is witnessing and makes the call. He notifies the offices of the Secretary of Defense, Duke Anderson, and The National Aeronautics and Space Administration (NASA) Director, Sharon Ross. He sets up the emergency group video call: Situation Room. After a few minutes, they start to join in the call. "This had better be good," one of them says. Oh, it is.

The Secretary informs them that he believes a "visitation" is underway at Area 51. I have the Secretary of the Department of Homeland Security joining the group. Secretary Matsuri patches the video call straight through to the President of the United States, as protocol dictates. His secretary Dianne Moore receives the video call and then notifies the President. The President picks up his phone. "Mr. President, you have an urgent video call from Area 51," she says. "It's all set up and ready to go." Thank you, Dianne. I need this room," he says. Everyone quickly grabs their things and leaves the oval office. He sits at his desk in front of the computer and joins the video call. Good day all, the suspense is killing me, what do we have, he asks? Secretary Matsuri explains to the group the situation and exactly what he believes is happening right now in

Nevada. No one else in the group says a word. "Are you sure?" he asks. "Yes sir," says Matsuri. "This is a visitation developing," he says, in an excited voice.

They can all see a fiery orange and red glow illuminating him and the background. The President commands, "I need you to stand by a moment." He mutes and removes his camera feed, then walks out from behind the Resolute desk. He goes over to a very old oil painting of George Washington which hangs over the mantle. Behind it, is a wall safe. He enters a combination of numbers, opens it, and pulls out a large, tattered old leather-bound book and a small device that looks like a cell phone.

The President proceeds to thumb through its pages, clearly looking for something specific. He finds it and dials a number into the keypad and waits for a few seconds. This is an Executive-level, Alpha, and Omega designation; the last visitation is here. This triggers notifications to the world powers that be, the small percentage of those who can afford to know things. He hops back onto the group video call. "Thank you for holding," he says. "Mr. President, we have a confirmed visitation happening at Area 51," the Secretary of the Department of Homeland Security, Thomas Mann says. "Contact has been made. The subject or subjects are already being tracked and communicating in the manner established from historical visits, from the fire cloud Sir." "From what we understand by reading the signals coming from the cloud, they have chosen you to receive the visitor," Secretary Matsuri says. "Shall I proceed, Sir?"

The President pauses for a moment to take in what he just heard. "Proceed," the President says. He sits forward on the edge of his chair and listens carefully. "Randolph, you and the other delegates from the G7, need to get here as soon as possible to greet them." It is official, the last visitation has begun. "I never thought this would happen in my administration," the President says. "This is big. It doesn't get any bigger Sir," Thomas says. "Ok, we'll leave from here in exactly one hour." "Understood Sir," they all say, almost in unison. The video call is dropped, and the video feeds go dark. The President stands to his feet and slowly looks up to the evening sky. He contacts Vice President Davis and advises him that he will be in charge for the next three days and to get Airforce One ready for departure. He alerts the other world representatives of the G7, staged in Westcoast region nine.

Kingston, Jamaica

Northeast of Havendale

July 9, 2021, 10:11 PM (ET)

Dr. Horace Michael Mitchel relaxes by a fire pit, while his three Grandchildren, Sammy age 9, Mona, age 12, and Josh, age 15 play nearby, in the backyard of his Jamaica retirement dream home, complete with a rock pool and four horses. Each summer he looked forward to spending quality time with his grandchildren, hidden away in the Blue Mountains, North of Havendale. This Summer the children have a subtle sadness hanging over them, like a dark cloud, as if a storm was coming. Their parents have been fighting like cats and dogs and even talk about getting a divorce if things don't change. The kids are glad to be away for the Summer and hope to forget all about the issues and drama they have at home. Sammy points up to the dark, star-filled sky. "I see the constellation, Orion." "Oh yeah, I see his sword and his belt with the three stars." "Hey, did you know that the three stars on his belt line up perfectly with the three Great Pyramids of Giza?" Dr. Mitchel asks. "I didn't know that," Sammy says. "I think I read that somewhere once. Do you think there's life out there, in all that?" "I know there is," Dr. Mitchel turns and looks at Sammy as he peers through the telescope. "You sound so sure of it, Pawpaw." "That's because I am," he says. "All that space out there, where does it end? Does it end or does it just keep going on and on forever?

Dr. Mitchel looks up, then down at Sammy. "What do you honestly think? What does your heart tell you when you see all those stars up there?" Sammy thinks for a moment. "I don't know. It makes me feel so small, but kind of safe too. I feel like we are being watched over." "Wow," Dr. Mitchel ponders over his very mature response. I think we know these things in our hearts, deep down we know that we are not alone. Dr. Mitchel walks over to the tiny fridge to get a drink. "There must be something out there. Yeah, I think so, there's got to be something out there in all that." A streak of light flashed across the sky. Sammy points to it and yells, "Look a shooting star Pawpaw!" Mona walks over to Sammy with her hands on her hips. "You don't even know what a shooting star is," declares Mona in an irritating voice. Sammy looks at Mona, then turns away. "I didn't think so. It's not a star, it's just a small piece of rock that hits the Earth's atmosphere stupid." Dr. Mitchel listens to the toxic conversation from the adjacent room.

"Mona, my darling, watch your words." Sammy looks at her and pokes his tongue out at her. "What have I always told you?," Dr. Mitchel asks. "Our words are the most powerful gift ever given to us by God," Mona says. "That's right."

The kids slowly turn toward the rich sound of Dr. Mitchel's wise old voice, one at a time knowing they are going to get an earful of wisdom. He slowly stands to his feet. "Don't ever get old," he says while massaging both knees. "You have to understand that our words have the power to change things for the good or for bad, bring positive or negative energy to any situation. Remember that your words can literally change the world. I wish more people understood that. It would be a much quieter world." Dr. Mitchel slowly walks over to his desk and sits down. "The President speaks and war is avoided, why? Because he spoke eloquently? No, because he spoke wisely, with carefully chosen words. He spoke words precise as a surgeon's knife. With a thought that became a word, everything you see in this world, everything in this universe was created." Dr. Mitchel pauses for a moment and just looks at his beautiful grandchildren. "I'm sorry," Mona says. "I forgot about what you always say to us. You always say a word is a living thing, a spirit, a seed that is planted in the soil of the mind." He smiles. "Wow, you do listen to me. Come here," he says. He hugs her tight. "Please use that big brain of yours for good." She laughs. "Ok," she says, "I will." "Now please go and apologize to your brother and mean it."

Mona goes over to Sammy and they begin to look at the stars together. "I wish your mother and father would remember to watch their words," he mutters to himself. Mona comes back over to Dr. Mitchel and asks, "can you do that thing you do with words, the illustration thing, please" giving him the "granddaughter" eyes? She realizes that her Pawpaw is getting old and this may just be the last time he does it for her. "Okay", he says with a huge smile on his face. Sammy and Joshua come back into the room and sit down to see him do it even though they have seen him do it hundreds of times. Dr. Mitchel stands up and begins his illustration of the power of words. "I'm about to say a word in a few moments. You don't know what that word is yet but in a few moments, you will. Any moment I will say the word. Here it comes. He puts both of his hands up like he always does and finally says the word. "Peace," he says, moving his hands from all the way right, to all the way left as he says it. "There it goes," he says, "you will never get that moment or that word back again. It's gone, swept away on the fabric of time." Mona silently mouths Dr. Mitchel's words as he says "swept away in the fabric of time" Or is it, he asks? The word I said a few fleeting moments ago lives on in that part of you that is eternal, to take root in your understanding and become a part of who you are. That's why we watch our words. "Peace," he says and finishes. They start to laugh and clap for him. "Thank you, thank you, Dr. Mitchel says as he takes a bow. "That never gets old," Josh says.

CHAPTER 2

THE UNEXPECTED

UNITED STATES AIRFORCE FACILITY, AREA 51,
(SECTOR 1) NEVADA "THE EXTRATERRESTRIAL
HIGHWAY", JULY 10, 2021, 12:23AM (PT)

Airforce One touches down and taxis the base runway. The President and his staff members take a short but speedy convoy ride to the location where the fiery cloud hovers over the base. They arrive and set up a security perimeter around the President, who anxiously awaits the visitor's appearance. Shortly after the President arrives, the delegates from the G7 (Group of Seven) nations do also. The entity or entities seemed to have been waiting for the representatives of this world to get there. The cloud of fire maintains a holding pattern above the base for several more hours. The Secretary of the Department of Homeland Security kindly asks for everyone's cell phones, iPads, Smartwatches, computers, and all recording devices as they are to be placed in a retrieval barrel. They reluctantly do so. The moment will only be captured by NASA historical media and the Space Force as current visitation protocol dictates.

They watch and wait. Just when the President and the others begin to feel like it's not going to happen or there has been some type of mistake, they hear an unearthly roaring tone coming from the cloud. The sound is ear-piercing and overwhelming to the senses. Everyone covers their ears. "What the hell is that?" the President shouts. "Is that them?" he asks. Space Force Soldiers stand ready to protect the President and his entourage. The sound abruptly stops as suddenly as it started. "This is it everyone; they have finally come back," the Secretary of the Department of Homeland Security says to one of his staff members standing beside him. He looks like a kid on Christmas day. "I've been waiting for this moment my entire career," he says. Delegates from every country chatter and write notes, while gazing up at the cloud. No futuristic spaceship, no flying saucers, just a fiery cloud has come down. The glow from the cloud turns the entire valley a reddish-orange.

The visitors finally emerge from the fire cloud. They step out of it right in front of them all. Fire from the cloud they came down in, covers their muscular bodies from head to toe. It quickly starts to dissipate and dissolve off of them as all three of them float down to the ground, burning the ground as it hits. They each stand about twelve feet tall and are fierce and intimidating to look at. The liquid fire continues to drip from their bodies, leaving no burns on them at all. Their clothing is not singed, nor disturbed in any way. They all look on in amazement and wonder at the visitors, who appear to all have dark skin. They could be easily mistaken for extraordinarily tall black men. Blinding bright light shines out of the back of their wooly, hair-covered heads, like a traditional halo you would see in one of those old oil paintings. "What the hell," Secretary Mann says. The group can't believe what they are seeing.

For all intents and purposes, the visitors are black and look like any other Negroid person, with slight variations. The President cautiously steps forward to greet the main one out front. He cannot believe his eyes. He gathers himself, clears his throat, and says, "I am the President of the United States of America. We welcome you back to your planet. We yield to your awesome power." Each person present, quickly bows down to one knee, as the visitors listen to the President's words. The one that appears to be the leader steps forward and says something in an unrecognizable tongue, clearly not of this world. He pauses for a moment, then speaks in perfect English, while the other two stand silently behind him like large statues. "I am Pherous, a Prince of Adimowa, the place that created you and everything in this world. It is written in the great books, preserved by the Highest." As he speaks, they can barely remain in a kneeling position and must lay flat on the ground. Some of them have even passed out. Wind and electromagnetic power can be tangibly felt in his voice.

The fear that grips them at that moment, is like the fear felt during a natural disaster, like a tornado or a hurricane. They do not doubt that they are in the presence of an awesome power rarely experienced by man. Realizing that the soundwaves of his voice will physically kill the frail humans, he consciously lowers it. "The Kingdoms of this world have always belonged to me," Pherous says. He points at the President of the United States and asks, "Are you the one in charge of my world?" "Not anymore," the President says. "Well-spoken," Pherous says. The rest of the foreign delegates look around, each of them hears Pherous speaking to them in their distinct language. They are amazed by this fact. He directs his attention to the U.S. President. "Did you know that this would happen one day?" "Yes," the President says. "The knowledge of you has been passed down throughout the ages and has been written in books, as you said. You were known by us to be the Vimanas in the four Hindu Vedas, the Chariots of God to the Jews and Christians, Ezekiel's wheel within a wheel. The Anunnaki to some. You are the ones that created men and placed them in Africa long ago.

We haven't experienced a visitation in over 40 years. Pherous laughs to the sky. One human day is that of a thousand years in my world. Pherous curiously looks over the diverse faces of the delegates and staff on the ground. He has never been to planet Earth and has never seen skin like that of the President up close before. "I find your appearance odd, "he says to the President. "Your blood shows through your skin, he says. He touches the President's face with his massive fingertips. The President feels the essence of Pherous' powerful aura, as he makes contact with his face. The President stumbles back and falls to the ground, unable to stand up. He regains his composure, but still can't believe what he is seeing or hearing. Pherous calmly walks over to him, towering over him. "Like many other messengers from the past, I have come here to deliver you a message," he says. "One of despair, of great calamity, and destruction, that will one day become one of hope and peace. The world that you once knew, is no longer. Your time to rule the Earth is over. Rise and look at me," Pherous says.

The President is helped to his feet by staff members and steps back to look into the face of this great being. The rest of the Space Force Staff and Soldiers remain face down on the ground, paralyzed with fear they have never felt before. "Why have you come here to the United States of America?" asks the President in a terrified voice. He hopes to demonstrate to the other nations present that America is still superior to them. Pherous looks at the group. "Have you not known the scriptures, you have been chosen by the Highest, I am here because the eagle is your symbol, and you are the last "Great Kingdom of the North" are you not? Do you not sit amongst the stars?" The President nods his head in the affirmative. The other nations now know their place. Your ancestors once served the many gods, such as Qos, a deity created to honor and worship my mother. This place here is also the gateway to your world from mine," referring to Nevada Area 51. He gently and almost lovingly takes the President's tiny man hand and says, "Come, let us reason together," Pherous says. They walk away from the others, then stop a few yards from the entrance of the bunker.

Pherous energizes his body with his thoughts and creates a large, pyramid-like structure made of light, that enlarges to surround him and the President. The rest of the President's cabinet and entourage can only watch, as the President is taken into it. The other two Adimowans remain there like sentries, standing watch and never saying one word. They are sailing into uncharted waters now. The rest of the group starts to muster and begins to stand up. They talk amongst themselves, having a thousand questions. The Secretary of Homeland Security turns to the Chairman of the Joint Chiefs of Staff, U.S. Army General Forester, and asks him, "Did you know that they would look like that?" The General shakes his head up and down, then says, "Yes, but color doesn't matter, right?" Thomas gives him a look of disgust and starts to walk away. "Wait," he says. Thomas stops and turns back. "This is the reality. Everything else up till now apparently has been theater. We inherited lies, my friend. We come from them. They made us; those are the facts," General Forester says. "At least now we know. Yeah, I guess so," Secretary Mann says. "Would it be ok with you if they were green, grey, or Nordic? General Forester asks. Thomas seems to have had his hopes set on something else.

Pherous looks over in their direction as if he can hear them, even though they're whispering. Thomas walks away visibly disturbed, an angry scowl on his face. The President sits, while Pherous stands over him, like a father standing over his child. "I came to you because I knew that you would be able to see what I am offering you and take full advantage of it. I'm offering you your New World Order. Isn't that what you want?" "Yes," the President says, with a confused look on his red face. "Forgive me, if I seem shocked, it's just that we knew that you were real, but this is the first time we see you with our own eyes." Pherous gives the President a look of love and even concern. "I'm not what you expected, am I?" The President shakes his head no. "I have come because The Ancient of Days is going to destroy your world by fire. He will use your inventions to do it. Everything you've ever loved or cared about will be gone. He plans to exterminate you, his creation that he sees, as completely gone astray. A clean slate.

He has done this in the past as recorded in your ancient writings. Last time by flood, how ironic to be by fire this time? Do you understand what I'm telling you?" "Yes, My Lord," the President says. "Don't ever call me that again," Pherous says with disdain. "That title does not belong to me." "Understand that no one in this world can do anything to stop what is coming," Pherous says. "But we have Space Force, our armed forces, and the other militaries of the world at our disposal. We even have nuclear weapons as options too." Pherous laughs out loud at the idea of a man fighting against the Highest. "You are, how do you say, like ants to him." Pherous puts his heavy arm around the tiny shoulders of the President. "I will protect you and show you how to preserve your right to live. Without my help, it will happen as it is written." "How can we fight that?" the President asks. "There has been rebellion before. Mistakes were made but not this time," Pherous says. The President stands up and thinks of his family and all his friends. "How?" the President asks. "I will complete what was started. I will make a new Earth, a better one." The President smiles. "When you do this for us, we will be forever in your debt." "Yes, you will be," Pherous says with a sinister smile.

He stands and towers over the President. "You will be like gods, you will be my people and I will be your God." "That sounds good to me," the President says. "Bow to me," Pherous says. "Pledge your faith and place your trust in me and I will save you and the families of Earth which you choose to save." The President knows exactly what Pherous means by that; he will be the one that selects who lives or dies. Pherous stares at the President then touches his forehead with his index finger. The President feels pressure in his head. Out of nowhere, his nose begins to bleed. "We have formed a covenant in blood," Pherous says, "which cannot be broken. Do you understand? "Yes I do," the President says. The back of Pherous' head starts to glow very brightly. The President stands to his feet, then slowly falls to his knees in complete and willful submission. "Open your mouth," Pherous says to the President. The President can't imagine what is about to happen next. He opens his mouth up wide, puts his head back, and closes his eyes tight. Pherous produces a small ball of bright light and places it into the President's opened mouth. "Swallow it," Pherous says. The President does. "What was that for?" the President asks. "I have put my seed in you." " Wait, what?" the President asks, with a very confused look on his face.

"I will give you a new world. We will create it together, in my image." "Why are you doing this for us?" the President asks. "We are not worthy." Pherous stops and turns away from the President. "Because your creator does nothing but hurts the ones I love; he has made so many orphans and widows of the children of Adimowa and the children of Earth. He causes you to experience so much pain and suffering in this world. He is a tyrant King!"

shouts Pherous. Concerned about the President, the Ghost recon team members that only report to the President ask, "Are you ok Mr. President?" "Yes, I'm fine! Don't come in here!" Pherous walks around the President as he sits in a type of chair. He gently strokes his face and hair. "I will help you but there is one who will try to stop us. It is written that he may fail if subdued in rusted iron chains for a season." "Why rusted metal chains, if I may ask?" says the President. "Everything in both Adimowa and Earth has properties that affect the matter around you." "I see, but I don't understand why anyone would want to stop us from saving the World. Who is it?" "In your year, 1979, this man was with a female royal. He knew her. She was sent here to help further the affairs of man, but she illegally opened herself up to him." "He had sex with her?" the President asks. "Yes. The visitation was authorized and sanctioned by the Highest, but she violated the laws of Adimowa by what she did. No mixing with your kind."

"We can assume that through the knowledge she gave him, he knows exactly how to get to the highest levels of Adimowa. He will attempt to have a council with Makal, the High Arc of Adimowa. He will alert him, then he will fortify his armies against us. I'm sure he will also look for Makia. That would all be very bad for us. The Arcs would find us, no matter where we went. The President thinks of the vastness of that statement. We must chain him up, not kill him until we have completed the task. "Why can't we kill him?" the President asks. "How is it you do not know this? You cannot kill him because your energy leaves your body and this world when you die. You never die, only change forms, under a new set of laws. Immediately, your life energy returns to the *River of Life*, in Adimowa, where the Arc sits above its entrance. He receives the souls. After they are collected, he is the one who weighs them in judgment for the Highest. That is the place where it is decided whether you stay or go." "Heaven or Hell," the President says. "Adimowa or outer darkness forever separated from the light of the highest," Pherous says. Only the Highest can bring you back from that. Most never return. Makal is also in charge of his armies. Killing him is not an option! "Why not?" the President asks, as if killing someone is standard practice. It was the way you were designed. It is written law in Adimowa. "I understand.

So, where is he?" the President asks. "I'm not exactly sure, but I do know that his energy is still here." "Ok, then that's easy, we'll send a team and locate him. It shouldn't be a problem; he won't be hard to find. It sounds like he's already on the federal grid," the President says. "Who is he?" the President asks. "He is Horus. Horace in your language." The President turns to his top General. "I think he's talking about Dr. Horace Michael Mitchel Sir," the General says. It must be him. He was already on our clean-up list. How many "Horaces" could there possibly be? "I remember him," the President says. "I never liked that guy, now I hate him," the President says. "Do not underestimate his capabilities," Pherous says. "He has been given much knowledge, more than anyone in this world, including you Mr. President," Pherous says. "He is protected by the knowledge he has been given." "But he's just one man," the President says. "He is one man that has prophecy on his side. That's what we are fighting against, so do exactly what I tell you." The President nods yes and says, You're the boss." Pherous removes the energy structure from around them. They both walk out of it hand in hand. "What else can we do for you," the President asks. "Just do your part and stay out of my way. I will do the rest." "Where do you want your new Earth to be?" Pherous asks. The President turns to his scientific advisers. "Well?" he asks. "Sir, there is a moon off the planet Jupiter that we believe is the best and most suitable to live on, given the right conditions. It's called Callisto," his adviser says. The second-largest moon of Jupiter. "Then

Callisto it shall be," Pherous says. "I will help the process along. Make ready your choices for the transition.

Everyone present, except Secretary Mann and a few others, begins to curiously surround Pherous. Most of them don't care that they look fierce and black, only that they are from another world. Some had no idea that we weren't alone in this universe. "May we touch you?" Secretary Matsuri asks for the group. Pherous and the two others give them a confused look, then they nod yes. "We don't bite," Pherous says. They all start to touch and grab their lower legs and thigh areas. After a few minutes of celestial bonding, the burning cloud of fire returns and hovers over them. "We must go now, but we will stay here on Earth until this is complete." All three of them re-enter the fire cloud vehicle and disappear. It's gone in a matter of milliseconds.

They all start to clap and cheer, still in complete shock and awe from what just happened. The President, his guards, his delegates, and the G7 leaders return home to ready their nominees to be "the chosen" and approved. Those who are expendable and nonessential personnel, just return to their regular duties and go on back to their lives. They are sworn to secrecy.

CHAPTER 3
SECRETS TO DIE FOR
KINGSTON JAMAICA, NORTHEAST OF HAVENDALE, JULY 13, 2021, 9:23 AM (ET)

Just after breakfast, Dr. Mitchel and the kids prepare to go on a short hike to a nearby reservoir. "That was the best breakfast you've ever made Pawpaw," Sammy says. "Thank you very much," Dr. Mitchel says. "We should take our fishing poles," Josh says. "Maybe we should," Dr. Mitchel says. "If you don't catch anything you don't eat tonight," he says. They laugh. "No way Pawpaw," Mona says, "I need to eat." "I like that idea," Sammy says, "earn your food." "Are we really going to eat the fish?" asks Mona. They all look at her. "Yes, we are," Josh says, "if we can catch them." Mona makes a yucky face. "I need to put that fire out before we leave; I don't want this place to burn down."

"What are you going to get me for Christmas?" Mona asks. "Now you know I don't celebrate Christmas, it's not Jesus' birthday, he says with a smile." "I know but why don't you?" The phone rings. "Saved by the bell," he says. "I better get that; it's probably your mother." Josh makes a face, as he thinks about his parents always fighting and arguing. "Why don't you have a Smartphone or watch yet Grandpa?," Mona yells. "Don't need one baby girl," he yells back, as he runs into his office to answer the phone. You wouldn't have to run to the phone, she mutters. They laugh at his old man ways. The kids continue to get ready. He's breathing very heavily and is out of breath from the short trot to his office just across. He pulls his heart medication from his breast pocket and then answers the phone.

"Hello, who is this?" No one answers back. "Shaggy's Rib Shack, may I take your order please," he jokes. He hears someone lightly breathing on the other end of the line. "Who is this?" he asks, in a serious tone. "Horace, this is Drew Slater. I hope that you remember me. We worked together on the thing in 1979." "Yes, of course, I do," Dr. Mitchel says. "Check your e-mails and you'll see one with the subject line that reads: "From the Place Where God Sits." Drew quickly hangs up. "Hello. Hello." Dr. Mitchel slowly hangs up the receiver then walks over to one of his bookshelves on the other side of the room. He takes his heart medication standing up, then sits down at the computer to check his e-mails.

From the place where God sits? God sits on his throne in Heaven. Adimowa! His eyes widen, as he looks through his e-mails. He sees one from "The Throne Dweller", thinking that it might be some hacker trying to give his computer a virus, but he opens it anyway. "Horace, they know what you did. They know that you have what they need. They are sending a team. Destroy all your research and any traces of it." Dr. Mitchel sees a scripture and verse from the apocryphal books of the Kings James Version of the Bible, II Esdras 9:1-2. Dr. Mitchel gets his little red copy of the Apocrypha off a nearby bookshelf, He finds the page and verse and reads it aloud, hoping to gain clues from it.

II Esdras 9:1-2 says, **He answered me then, and said, measure thou the time diligently in itself: and thou seest part of the signs past, which I have told thee before wherein the Highest will begin to visit the world which he made.** Dr. Mitchel gasps and stops reading aloud. "Visitations," he mouths to himself. She said that it would happen but she never told me my babies would be here when it did. "The children Makia," he says out loud. Dr. Mitchel looks down at the date of the e-mail, which is dated three days ago, July 10, 2021. "Oh my God," he says, with a look of complete fright on his face. He looks outside at his grandchildren sitting in lawn chairs, then he runs over to the fireplace and douses the flames with a pitcher of water. White steam fills the air.

He begins to grab papers from his desk and shreds them in the shredder. I shouldn't have put the fire out, he thinks to himself. A now extremely paranoid Dr. Mitchel peeks out of the front blinds, thinking someone might already be here. He runs to lock the front door and he then proceeds to slam shut several open windows. He starts to realize that the kind of people that are coming will not be stopped by locks or latches. He knows this because he used to be one of those kinds of people, the highly trained, skilled, and motivated. He knows that he has a real problem on his hands and needs to figure out something pretty quick. But what, he thinks to himself? He grabs his old leather journal from his desk drawer and throws it in the small trash can right next to his desk. He quickly picks it right back out and thinks "that was dumb." He is torn between destroying his journal or using what's inside. "Fight or flight syndrome" kicks in and he must decide now because the clock is ticking. "I'm not ready to do this," he thinks. "I can't do this," he says.

CHAPTER 4

FULL DISCLOSURE

KINGSTON JAMAICA NORTHEAST OF
HAVENDALE, JULY 13, 2021 9:37 AM (ET)

The kids are each on their gaming devices while they patiently wait for Dr. Mitchel to finish up with his call. Sammy stops playing for a second and notices that something is very strange with Dr. Mitchel. He sees him running around his office like he caught on fire or something. He runs back and forth like a chicken with its head cut off. "Look you guys, what's wrong with Pawpaw?" Sammy says. "What happened?" Mona asks. Josh doesn't really even look up and just keeps playing his game. He's just running around his office like something's wrong. Mona and Josh both turn back to playing their games. "It was probably mom on the phone complaining about dad again, or they're having another fight or something like that," Mona says. "I don't know," Sammy says, "but something's not right." "Yeah, I just wish they would get a divorce already and just get it over with," Josh says, clearly irritated. "What! Are you crazy?" Mona asks. "Yeah, don't ever say that," Sammy says. "Why would you say that Josh?" Mona asks. "We don't want them to get a divorce," Sammy says. "I know, but sometimes these things just happen. You'll understand it better when you get older," Josh says. "I don't know what I'd do if they got a divorce," Sammy says. "I know little brother, this sucks. We shouldn't even have to be thinking about this kind of stuff." "Yeah," Mona says. "Louis is your dad, not mine."

Dr. Mitchel struggles to deal with what he just read. Suddenly, he knows exactly what he must do. He calls out for the kids and turns around toward the backyard; they are already standing in the doorway, silent and looking up at him. "What's wrong?" Josh says. Did you take your medication? Yeah, what's going on? Should I call mom and dad?" Mona says. "He's not my dad," Josh says. Are you having a heart attack?" he asks. "I'm calling nine one one." Dr. Mitchel looks at the kids but doesn't say anything at first. No, I'm fine. He holds his leather journal tightly in his hand, thinking of how to put it to the kids. "Was it that call you just got?" Josh asks. "Was it them again?" "They're getting a divorce, aren't they?" says Mona. The kids wait for Dr. Mitchel to give them an answer. He finally does. "No, it wasn't them," Dr. Mitchel says. Stop talking! They stop and focus on the next thing he will say. "We don't have much time," he says. "Follow me." The kids hurry behind him, wondering what's going on that has Dr. Mitchel so rattled. "Go get the backpacks that I made for you, the *Bug Out Bags*." Dr. Mitchel runs out of the room, leaving them standing there dumbfounded. "Emergency," says Sammy, is this real?" he asks. "Yes, go," yells Josh, looking at his younger brother and sister. They run to the nearby closet and grab their bags. "We're going to be ok," Dr. Mitchel says. "Pawpaw you're scaring me," Mona says, as he quickly leads them down the stairs into the lower parking garage area and out into the backyard.

He stops and pulls them in close. "We need to get the horses ready and head out." "Out where?" asked Mona. "What is it, what's up?" "I never told you, but I worked on some government projects that were off the books." "What does that mean?" Josh asks. "It means that I have a lot of information someone either wants or wants to get rid of." He turns his back away from them, then turns back toward them and pauses for a moment. I'm thinking they want to scrub. That's a full sweep of the program and everyone that worked on it. Their eyes light up as they listen intently. I'm breaking an oath by what I'm about to tell you. Dr. Mitchel pauses a moment to take in the gravity of a secret that he has held for years. "We had a visit." "Visit? Visit from what?" Josh asks. Dr. Mitchel points to the sky. "UFOs?" Mona asks. Sammy quickly turns his head around looking at everyone in the room. "UFOs? Yes, you can say that we were visited by UFOs, they would be considered unidentified flying objects. But really it was a visitation from our creators. These beings are not really aliens at all. When you look at them, they are just like us, only a whole lot bigger, stronger and smarter.

They have all the basic parts that we do. They all have dark skin and even wooly hair just like us." "The aliens are black?" Mona asks. I don't know if they're black, I just know that they look like us and we came from them. So, yeah I guess so. They are definitely not little green men from Mars or alien grays you see in movies and television." "What about E.T?" Josh jokes, "He was black too, right? They all start to laugh until their sides hurt. "This is a prank right?" Josh asks. Dr. Mitchel stares at them for a moment with a serious face. "No Joshua, this is real, as real as it gets. This is reality. These beings are the ones who made us, made everything." "Wait a minute, you're serious?" Sammy says. "No, he's joking," Mona says. Thousands of years ago they were the ones that placed us in Africa. It wasn't called Africa then but it's the same place. They gave man laws to keep, to see if we would show our love for the Highest by keeping them. We obviously didn't keep his laws and were punished. We were supposed to live forever. The curse of death was placed on man, and we were forced to leave the garden. "Hold up, that sounds just like the Bible stories we read in Sunday school," Josh says.

"The truth is, the Bible is historically accurate. Adim was formed from the dark rich soil of Africa; Evi was taken out of the side of Adim, just like it says. You see, what we see as amazing, and miracles down here are completely normal in Adimowa." The kids' mouths all hang open for the next few seconds, but Mona still doesn't believe what she is hearing. "I was chosen to study one of them for three years. During that time, she gave me things that benefited the world in many ways. It's all here. He holds up his journal. I'm willing to bet the farm that this is what they want and are coming for. In this book are things that would blow your mind, things you could never even imagine. The call I got, and the warning email too, said that they know where I am, and they are coming. That puts you in danger and I can't have that. We need to go now." "Go where?" Sammy asks. "We'll take the horses out to a special spot on the edge of the property. I keep everything there for safekeeping." They take off on their horses toward the hidden location. "Get up now!" they all yell. Dirt and leaves fly behind them as they ride hard. They arrive at the spot in minutes. Woah! The horses come to a stop. "Here it is," Dr. Mitchel says. They get down from their horses and tie them to a tree about thirty feet away from the spot. It's well hidden. Dr. Mitchel begins to stomp his feet on the ground to find exactly where the entrance is. He hears a hollow sound. "Here it is," he says.

He uncovers the dirt and leaves to reveal a large wooden door. He takes his keys from his pocket unlocks the large padlock and opens it. "Come on," he says, as he leads them into the dark space and down a spiral staircase. Halfway down is a large light switch on the wall. He flips it and the lights flicker on. Dr. Mitchel reveals his hidden laboratory. "Well, what do you think?" The kids can't believe their eyes. "Wow, this is awesome," said Mona. He seals the door behind them. "This is my sanctuary, my man cave, my life's work. Everything is here. I have been preparing for this since I learned of what would eventually happen to Earth. This is our insurance policy. I'm glad that you guys are here in Jamaica with me." "No way! This is so cool. You made all this?" Sammy says. "No way you did this!" Josh exclaimed. "Every Summer we are out here." "All this time, this has been here? Why didn't you show us this before?"

Josh asks. "I know it's a lot for you to understand right now, but we are at the beginning of something huge. Something that I believe will affect the entire planet. Dr. Mitchel walks over to the three computer consoles and powers them up. The lights in the lab flicker on and off. Once the computers fully boot up, he begins to look at the three screens. "I'm sorry I hid this from you, but I had to. Some very bad dudes are on their way here right now. I'm pretty sure they want to take me out because of what I know. I have something they want." "What do you have?" asks Mona. "I'm not sure exactly," he says. "Have a seat," Dr. Mitchel says.

The kids sit down on his old comfy sofa and almost sink into it. "I wanted to tell you about this, but you guys were so young. As I said, I was sworn to secrecy, I really couldn't tell anyone. I think that It could be that I'm the last one with any knowledge of Project Mogul. "Project what?" Josh asks. He paces the floor as he speaks and walks over to his bookshelf. This will be the *last visitation*. The President will be the chosen one this time. That's here too, referring to his old journal. "Pawpaw what are you talking about?" Sammy asks. "I know this is a lot for you to process but try to forget about everything that you ever thought you knew or learned. You need to get the full story to understand how deep this goes." Dr. Mitchel (Pawpaw) goes on to describe how NASA scientists and the U.S. Military staged the retrieval of a fake alien body in Roswell, New Mexico in 1947. They ran that story in newspapers around the world. "Fake news," Mona says. "Yes! It was all lies. All a big cover-up from start to finish. There were no crashed alien ships or bodies retrieved; it wasn't even a weather balloon. That year there was a visitation I'm told. "In 1947?" Mona asks. "Yes," Dr. Mitchel says. I was selected to receive the visitation that took place in 1979, almost thirty years later. The secret project was called "Mogul".

The U.S. Military started the little green and alien gray narrative, and it worked. After that people were made out to be crazy if they even believed UFOs, let alone saw one and reported it. They became the tinfoil hat-wearing types. You see, they had the world, looking for the Hollywood version of spacemen, instead of what it really was. It's like giving someone the wrong spot to dig for a treasure that isn't even there. All of this was done to redirect the world's collective consciousness. The visitation in "79," Project Mogul, was special. I was literally the one chosen by Adimowa to receive the visitation. That's the way it worked. They always choose the host from the scientific branch of the most powerful government of the current age. Makia said that I was chosen for a purpose that would be revealed to me in my lifetime. She said it was already written in the ancient prophecies to happen this way. I was given a team of scientists that worked with me, but they were not allowed to interact directly with the fire cloud. This was a controlled environment. I spent exactly three years with her in a biosphere. I was removed from society, devoted solely to the project. I was completely in charge and had no one to answer to. Her name was Makia, a royal and only daughter of Makal the Arc of Adimowa.

She was extremely intelligent and spoke all the languages of Earth fluently. She told me that she was from a place called Adimowa or what we refer to as Heaven. I knew of her kind and the secret relationship between her kind and Man. I learned everything about her and received all the new tech for the Earth. She gave us microwaves, cellular phone technology, Kevlar in bulletproof vests, Bluetooth, even the cure for cancer believe it or not. "But people still get cancer and die," Mona said. "I know, but I gave them the cure she gave me. The Centers for Disease Control (CDC) has it, and so does the World Health Organization, but they decided to never release it." "Why would they do that?" Josh asked. "I don't know, I guess to keep the population down. I was so naïve at first, giving her world history books, Almanacs, and then Encyclopedias to read, thinking that this would assist her in some way." "What are those?" Mona asked. They were the Google of back then, the sources we got information from." "Oh", they said together. She already knew everything, I mean everything about our history like she had been studying us, watching us from afar. She showed me so many things about the universe, exceptions to the laws of physics, how to travel through black holes, and even manipulate gravity. Stuff like that. The kids have no clue as to what he just said.

I later learned that the only purpose for this was to give continuous visual honor to the Highest. It was a reminder built into their bodies of who the creator was and where the light comes from. She was extremely beautiful, I mean beautiful and perfect in every way. She looked like one of those Egyptian goddess statues. The kids all see that look of love in his eyes as he tells them all about her. I was shocked at how beautiful she was. I wrote everything here in my leather journal." "Can I see what you wrote about Makia in your book Pawpaw?" Mona asks. "Yeah, let's have a look," Josh says. "No, you cannot," he says. "Too personal." "Shush, let him keep going," Sammy says. "Then what happened?" "There were many things that I never even turned over to the Government. We spent every waking moment together. She trusted me and shared ancient and forgotten knowledge with me. I would partially report my findings once a year to my superiors. They were happy if I kept giving them new tech.

I think that the government at that time must have supposed that she was just a puff of smoke in the laboratory. She gave us so much but the one thing they were most interested in was knowledge regarding terraforming." "What's that?" Josh asks. "Terraforming is the modification of the atmosphere, temperature, surface, topography, or ecology of a moon or planet. Basically, it's changing let's say Mars into the Earth. People could live there." "I told her that this kind of information could be very dangerous if it got into the wrong hands. I held back that information. I didn't give them all the technology they needed to make it happen. They will get it soon though, from this last visitation, I'm sure of it. I can see why they kept this secret. It would spark global fear and chaos. It would blow the idea of all the world's religions right out of the water. There's only one religion anyway, which isn't a religion at all. It's reality.

The government needed religions to stay in place in order to create and maintain division and confusion. It's crazy huh? She told me the origin of everything." "That's why you're so smart," Mona said. "I don't know about that," Dr. Mitchel said. They can't believe what they are hearing, it's all so incredible. At that moment, an alarm sounds inside the laboratory. "Oh no, they're here!" Dr. Mitchel says. Too soon, we're not ready at all. I need to start up the flight preparations sequence for the Issachar. "The Issachar?" kids all wonder. It takes about thirty minutes to boot up for takeoff. "We're not going to make it," he says. "Ship, what ship?" Mona asks. The kids look up at Dr. Mitchel for what to do next, while wondering the whole time, about this ship he was talking about. Dr. Mitchel enters a seven-digit code into the computer which reveals the Issachar.

He runs frantically pushing buttons and grabbing things they will need for the trip. The computer starts a chain reaction that makes the room begin to shift and move. The laboratory floor underneath them starts to slowly open, revealing the craft. It is designed with Adimowan, highly advanced technology that this world has never seen before. "Wow!" they all say. "You know how to fly that thing? I've never seen anything like this. Are we going in that?" Josh asks. "Not if they get here before the flight preparations are made." "You know how to fly that thang?" Mona asked. "Yep. I

built it, but I've never actually flown it before."
"That isn't from NASA, is it?" Sammy says. "No,
it isn't, like I said, I built it, with Makia's help of
course and I do have a degree in
Astrodynamics from MIT. You guys get in and
get ready to go." The kids still had very
worried looks on their faces as they boarded it.

Kingston Jamaica

Northeast of Havendale

July 13, 2021, 9:45 AM (ET)

"We're here," the Team Leader, Marine
Sergeant Jeff Aragon says into his earpiece
radio. A four-man fire team fast ropes from
their little bird helicopter to the ground. They
silently move toward Dr. Mitchel's cabin.
Unknowingly, they trip several motion
detectors along the way. They arrive at the
front of the cabin and prepare to make their
tactical entry. "Ok, listen up. We've been
ordered to take him alive so no mistakes." "Tap
up," Jeff whispers. With the breacher and
ballistic shield upfront, they toss a concussion

grenade through the front window. Glass breaks. Boom! They rip the door right from its hinges. "Go, go, go!" Jeff yells. The team does a crisscross entry pattern and quickly clears the house. "All clear! Check the backyard," Jeff says. Half of the team heads out to the backyard. They see that the stables are empty. A team member radios to Jeff. Intel said that they had four horses, right?" "That's what it said," Jeff says. "Sir, the horses are gone," one of them replies. Damn. He radios the eye in the sky. "They're not here but I think they're close. Do a sweep of the area. It looks like they left in a hurry." The helicopter uses its infrared lens to search for heat signatures in the area. "I got something; I have four warm bodies three clicks East of the Cabin. They are all hunkered together in one area sir." "That's them, let's go," Jeff says. The team regroups and heads out to the new location.

CHAPTER 5

"BROTHA" NAMED HORACE?

WASHINGTON, D.C. MARINE MEMORIAL CIRCLE, JULY 13, 2021 9:48 AM (ET)

F ire Chief, Greg Miller, wraps up his press conference regarding a tragic laboratory fire. He is surrounded by no less than ten city officials, some in suits and others in full dress police and fire uniforms. "Certainly, today our prayers, well-wishes, and thoughts go out to the Fischer family. At this time there doesn't seem to be any foul play involved and it appears to clearly be an accident, but the investigation is obviously still open and ongoing. This will have to be the last question." The Chief points over to news reporter and former Chicago Bears cheerleader, Zoe Baker. "You there," he says. She gathers her thoughts, "Thank you, Chief

Miller, Zoe Baker, KNRA News. I know that this is an ongoing and active investigation but is it true that Doctor Fischer worked for the United States Government and is one of several NASA Scientists to disappear or die recently?"

Chief Miller looks around as if to get permission to answer the question. "I think it's important to let you know right now that just before coming out here tonight, we were advised by the powers that be that going forward the FBI would be taking complete control of this investigation." The reporters begin to circle like sharks around him and clamor amongst each other, begging for him to elaborate. "As far as Doctor Fischer being a part of some top-secret project in the desert, that would be news to all of us." Some of the other officials look at him with widened eyes as if to say *shut the hell up*. "Our preliminary investigation shows that Dr. Fischer had a gas leak in his home's old-style gas furnace in his basement. Chances are it was not serviced properly and developed a leak. There's no big mystery here, no smoking gun.

It's simple, gas and fire don't mix. Thank you, no further questions." Chief Miller hurries away from the podium as the cameras click and flash, The throng of reporters try to ask him just one more question as he leaves. All the reporters and journalists run to their news fans and attempt to be the first to get the story out. "Zoe!" Alfonso shouts, "you ok?" he asks. We need to get this out before they do! For some unknown reason, Zoe zones out for a few moments. She seems somewhere else at that moment. She comes back from wherever she went in her mind. "Yes, I'm fine," she says, snapping back into reality. Zoe and her cameraman Alfonso Nazario get the story out just in time for the 11 o'clock news.

"This is Zoe Baker reporting live, back to you in the studio," she says. Alfonso gives the all-clear signal to Zoe and turns off the camera light. "Great job," Alfonso says. "Thanks. I wonder why he said top-secret project in the desert. I never asked bout anything top-secret," Zoe says. I think they know more than they're letting on. "What makes you say that Alfonso asks?" "This is way too strange to be a coincidence, and why is the Federal Bureau of Investigations coming in to take over? I mean, Doctor Fischer was a great scientist and all, but a federal case out of this." "No, it's weird," Alfonso says. What makes this case so important to them? It just doesn't fit. Certain competitive reporters are still standing around trying to see what else they can get. Zoe leans over to Alfonso and whispers in his ear, "I found out that there's only one scientist left on the East coast that hasn't gone missing yet." "I'm all ears," Alfonso says. "You do have some big ears," Zoe says jokingly.

"A doctor named Horace Mitchel." "Horace?" Alfonso laughs at the rare and funny name. "I like that name; you don't hear it a lot," Zoe says. "We need to get to him before they do, to find out why they are doing this sweep. The only problem is he retired and moved out of the country. I heard he's in Jamaica somewhere." "He's a *brotha*"?" Alfonso asks. "Yep. I had a picture in my mind of Bill Nye the science guy. How do you know that?" Zoe smiles at Alfonso while continuing to think about what to do next. I do what it takes, so don't ever cross me. "Oh, believe me, babe, I won't unless I'm ready to head for the border. I'm just glad you're on our side. Damn, I would hate to have you on my ass." Alfonso stops to think about the visuals of what he just said and smiles like a teenage boy. Zoe doesn't smile back at Alfonso as she pulls out her phone and gets ready to make a call.

"Horace, where are you?" she says. We need to find this guy before he comes up missing. I'm sure he knows exactly what's going on. "What do you need me to do?" Alfonso asks. "Try to find as much information as we can on this guy. I know that you have resources too." "A few," Alfonso says. I'll reach out to some of my boys over at the Blackstone and see if they can run him. "Perfect," Zoe says. I'll work my end too. Alfonso gets another mental picture of that and smiles. "You are a pervert, Alfonso." "What?" he asks, with a devilish smile on his face. "I didn't say anything." "No, but I can almost see you thinking. Let's go." "Hey, you ok?" Alfonso asks. "I'm good," she says, "thanks for asking." Alfonso knows her very well and can tell that she's been a little off lately. Women, he thinks to himself.

They both get in their cars and drive away. After driving about a block or two, they both end up stopping at the same red light and looking over at each other. Zoe just stares forward still in a daydream. Alfonso rolls his window down and Zoe finally notices him and does too. "You ok lady," Alfonso asks? Zoe just smiles at him and then slowly rolls up her window. That's cold Zoe, Alfonso says to her closed window. Zoe gives him a thumbs-up as the light turns green. She points up to the green light and speeds away in her Porsche SUV, leaving Alfonso sitting at the light. The car behind him gives a rude extended honk.

"Ok, ok he says, just drive over me why don't you," he says sarcastically while looking in his rearview mirror as he drives away.

CHAPTER 6

STILL A "DEVIL DOG"

KINGSTON JAMAICA, NORTHEAST OF HAVENDALE, JULY 13, 2021 10:10 AM (ET)

The ship still isn't quite ready to go. "She will be ready for takeoff hopefully in five minutes," Dr. Mitchel says. "It won't be long before they figure out where we are. You hear that?" Dr. Mitchel asks. What?" Sammy asks. Dr. Mitchel hears the distinct sound of a military-grade helicopter hovering at a high altitude. "Oh Shit, sorry. They found us." "Oh shit," Josh says. "Don't say that," Dr. Mitchel says. "Sorry. We got to go now!" Dr. Mitchel yells. He bends over the computer console and strikes the "enter" key several times. "Come on," he says. Sammy and Mona almost run in circles, bumping into each other. "Here we go," Dr. Mitchel says. The ceiling above them begins to open slowly. The jungle covering the floor above them widens to reveal the laboratory below. The Issachar is just about ready for takeoff. Dr. Mitchel sees the helicopter hovering high in the sky over them. "They must have a ground team too," he says. "Come on!" he yells. They turn to run back into the Issachar but power in the Lab suddenly goes out.

Mona and Josh use the flashlights on their cellphones to light the way. Everything shuts down. "What the? No, no, no," Dr. Mitchel cries! I think we blew the breaker, or somebody cut the power. He remembers that the breaker box is located above them, about ten feet from the laboratory hatch door. I need to go up there and reset the breaker. "No," Mona says, "they'll hurt you Pawpaw! Just as he decides to go up, he hears a man whistle. Standing above them is the search team with their weapons at the low ready. "Hello there," Jeff says. "Please don't hurt the children," Dr. Mitchel says. "They don't know anything." "All of you come up the stairs and please don't try to be a hero," Jeff says. "We don't want to hurt anyone." Our orders were to take you quietly. The kids breathe a sigh of relief. Dr. Mitchel secretly grabs a small smooth stone from his desk and conceals it. Dr. Mitchel and the kids turn and start to climb the spiral staircase back up to the jungle surface. When they reach the top, Dr. Mitchel fakes being completely exhausted and falls to the ground. While on the ground he picks a handful of a local herb that he recognizes and holds some in his closed fist. The team removes their gas masks. "Are you Ok Old Man? Missed some PT (Physical Training) did we," Jeff says. The team all laughs at the tired old man. "This old guy used to be a "Devil Dog." He was MSOR (Marine Special Operations Regiment) a Raider," Jeff says. "What?" a team member says. This old guy?" The team is impressed but very disrespectful. "Yeah, did one tour in Nam too. Joined when he was just seventeen. Now that, I respect. He did twelve years on the Constabulary Force right here in Jamaica. Slumming, I guess."

They all laugh. "Fellas, you're looking at a genuine, made in the USA, badass. Hey Jeff, that's you in ten years." They all laugh. Jeff doesn't. "Ok, here's the deal, you're going to come with us now and you're going to do that quietly. Do you understand?" asks Jeff. "Relax GI Joe, I'm just an old man and these are just kids. We'll do whatever you say." Dr. Mitchel has had his hands behind his back the whole time Jeff and others have been talking. They don't notice that he has been grinding the jungle herb into a fine red powder. Just then Jeff asks, "What do you have in your hands?" "Nothing, just this," Dr. Mitchel says. He blows the red powder right in the faces of the team. They all stumble back with their eyes wide open and surprised. Instantly they lose all control of their body and collapse to the ground but remain fully conscious. Dr. Mitchel and the kids calmly walk over to the team and look down at them. He slowly goes down to one knee and gives Jeff a couple of light slaps to his cheek. "Wow, I bet you didn't see that coming." The team can still hear and see everything even though they are completely paralyzed. Dr. Mitchel takes their equipment and throws it into the lab below. Ok, here's the deal Jeff, you and your team will be like this for about three hours, so get comfortable. After that, you'll slowly regain some motor functions. You will be soiled. Sorry about that. The kids laugh under their breath.

Jeff and his team can only look up in disbelief. Dr. Mitchel resets the breaker, and the power is restored. They run back down the spiral staircase. He re-enters the takeoff sequence into the computer to get the Issachar ready. Most of the program was still running, saved when the power went off, so it won't be long now. The ship is finally ready for takeoff. I'll start the flight sequence now! Let's get the heck out of here before the pilot figures out that the team is out of the game. They all get into their seats and buckle themselves in. The Issachar starts its internal systems, readying itself for takeoff. "You guys ready to do this?" Dr. Mitchel asks. They all look wide-eyed and frightened but give the thumbs up. He does a final equipment check then hops into the captain's chair. Joshua is his co-pilot.

The ship begins to shake and move in a vertical take-off, ascending through the large opening above the laboratory. They start to climb over the still paralyzed bodies of the black ops team. They can only look in wonder. "I don't hear an engine," Sammy says. That's because it's made with anti-gravity propulsion. It's out of this world. In a blink of an eye, they thrust away. Inside the laboratory, there is a computer screen that reads 2 hours and 15 minutes and counting. Dr. Mitchel has set a timer that will detonate placed explosives in the lab, right about the same time the team starts to revive. In the lab, there is still sensitive information that he does not want them to get. They start to come around. "Get up you idiots," Jeff says. They are soiled, just like Dr. Mitchel said they would be. "Damn, this sucks," Jeff says. Very wobbly-legged they stand and begin to recuperate. They have no way to communicate so Jeff tells one of his men to build a fire. "Damn not even a flare. "We need to find them, he's going to be pissed, Jeff says. Maybe the little bird will see the fire." One of his men strikes the flintstone several times, then starts a small fire. Proud of himself he says, "we have fire." As he happily stands and turns toward Jeff, the laboratory beneath their feet blows them sky-high. Boom! They all fall back flat onto the ground. They are scorched from the blast. Jeff pats his clothing, quickly putting out the fire. "I really hate that Raider," Jeff says, with a disgusted look on his face. The helicopter Pilot sees the smoke and fire billowing into the air. The pilot swings back around and sees Jeff and the others waving. He searches for an open area to land.

Inside the Issachar Spacecraft
Over the Atlantic Ocean

July 13, 2021, 1:10 PM (ET)

"This is so cool!" Sammy says. "I'm going to throw up," Mona says. "You better not," Josh says. "Oh my God! Everyone just breath and try to stay calm. You can do this. It's just like a video game," Dr. Mitchel says. The Issachar goes subsonic as it streaks across the sky into the stratosphere. It breaks the sound barrier causing a sonic boom. Exactly Sixty miles above Earth they reach outer space and cruise at an altitude of about 47,000 feet. They are completely undetected by any radar system, in stealth mode. We're safe now. "How you guys doing?" "How are you doing?" Josh asks. This thing moves. The ship starts to level off and begins to orbit the planet. "You can take off your seatbelts and walk around if you want." The kids walk to the window and see space, with all the trimmings. "We are so high up," Sammy says. They also gave us the technology to terraform and that's what they are doing right now. Makia told me it would happen this way. "Terraforming is the process of making an uninhabitable planet or moon habitable, able to sustain human and plant life. It's turning, let's say Mars into Earth. With the technology that the world has today, terraforming is only a hypothetical, a dream." "They are building a new Earth," Josh says. "Exactly Joshua! They plan to destroy the world before God does. They want to start over and colonize the planet with only selected rich and powerful people of the world. You see, their plan wouldn't include regular people like us. We would be left behind. These men coming, tells me that everything Makia told me was true. "Where are we going?" Mona says. "Where we can figure out our next move. There are a couple of places on this Earth visiting Adimowans would hide out, places that were so hard to get to.

There was Atlantis near Bermuda, places in Siberia, Russia, and one in Antarctica. "Wait a minute, the real Atlantis?" Josh asks. "Yes.

Ancient Atlantis was destroyed thousands of years ago, but I know where it is. They would hide there underwater. "Are we going there?" asks Mona. "Nope, we're going to Antarctica. It's still a place where electromagnetic activity disrupts and interferes. There are a few research stations down there but not much else. You see, that's why the Antarctic Treaty System of 1959 led by then-President Dwight D. Eisenhower was set up. The treaty says that there are to be no military operations, no mining of natural resources, and no nuclear weapons testing. Even then they knew what it was used for. The location of one of the four entrance points went viral in 2006, but that was made to be a hoax, calling it a crevasse. A natural crack that occurs on the surface of ice or glaciers. That's crazy because that just doesn't happen in the mountains.

We'll be safe there. It should be completely abandoned. The knowledge of this place is above top secret. They reduce their speed and lower their altitude as they get near Antarctica.

Not long after that, an alarm goes off in the cockpit. A radio transmission comes through the radio. "This is the United Nations Airforce. You are flying an unregistered aircraft, in U.N. airspace. Land your craft now or you will be shot down." "They are not supposed to be here," Dr. Mitchel says. The F-22 has the Issachar in its sights. "Oh no, they just locked on to us," Dr. Mitchel says. He tries to shake the F-22 and deploys anti-missile flares. "Get strapped in." The kids get seated and secure their shoulder harnesses. "Hold on to something," he says. Then he presses a button. The Issachar goes into hyper speed and bolts straight into the side of the mountain hole and then disappears in the darkness. "Wow, what was that?" the pilot asks. "I have no idea," says the Co-Pilot. "I think that was a UFO." "Don't even say that." "Yea, you're right." The pilot radios in the report of the incident. Base command, it was just a big nothing up here, we're RTB! "Roger that," the base command says.

The F-22 Pilots return to base with a story they will never tell. The Issachar traverses the cave heading to the Adimowan outpost. The cavern is so massive that the Issachar looks tiny in comparison. I see something. Is that it?" Mona asks. The walls around them are made of diamonds or quartz. "I see the platform there," Dr. Mitchel says. "It certainly looks abandoned." The ship comes to rest on the flight docking area that's still there. The facility doesn't look old at all but is still magnificent. The architecture is like nothing on Earth. The ship powers down but remains on standby. You better put those cold-weather suits on before we go out there. The Issachar can become completely luminescent as it provides light to the cave in every direction. The brightness shines through the crystals all over the cave, which is quite breathtaking. They walk out together. "Oh my God," Mona says, "It's beautiful." "Wow. Cold," Sammy says. Steam comes from their mouths as they speak.

"Ok, let's make sure we're alone here." They all walk up to the door.

Dr. Mitchel places his hand on a futuristic reader and then says words written in his book. The door slides open with a *whoosh*. They go inside and get somewhat comfortable. Makia told me of all our ancient forgotten history and about places like this. "You keep saying Heaven Grandpa," Josh says. "So, God is real?" "Yes. Makia said she had never seen the Highest. No one has ever seen him and lived." "So, let me get this straight, everybody gets to go to Heaven?" Mona asks. "Yes and no. Everybody returns to the source. Our spirit returns to Adimowa. That part of you that's looking at me now, behind your eyes, that part returns to the River of Life. Our bodies are just containers. After that I don't know what happens," Dr. Mitchel says. "There are so many things that we don't understand. Ok, get some food in you and then try to rest if you can. I got a feeling we're going to need it." He sets down his large cooler pack and opens it. They grab food from it and then begin to explore. "You guys stay together and don't go too far." "I got to go to the bathroom," Mona says. "Me too," Sammy says. "Wait, here, put these on." He gives them all communicators, in case they need them. "Try to find a restroom, if there is one. If not, just make one." Mona makes a crunch face. Dr. Mitchel goes over to the table to think about what to do next.

The kids go back into the ship and the galley. They sit at the table. "I miss mom and dad," Sammy says. "They don't miss you," Josh says. "They're too busy fighting." "I bet if they knew what was going on they would stop fighting," Mona says. "Maybe, maybe not," Josh says. "They would see what's important," Sammy said. "You think they fight because of us?" Mona asks. "No. They just hate each other's ways, that's all. It has nothing to do with us." "They just look so happy in their wedding picture in the living room," Mona says. "Everyone looks happy on their wedding day. Stop trying to figure it out, I did. Listen. We have bigger things to worry about now. Look around you. Everything will be different from now on if what Pawpaw says is true," Josh said. "I bet they are starting to wonder where we are," Sammy says. If they only knew.

"Pawpaw are you from Adimowa too?" asks Mona. "No, I'm not. We are all descendants of Adimowa for sure. "How do you know that?" "I was very much in love once." "Makia, huh?" Mona asks. "Yes. She was the most beautiful, indescribable creature I had ever seen. She was my one." "That is so sweet," Mona says. "She had light, like a halo coming from around her head. That's where they get the idea of a halo in those old pictures. I was drawn to her. She was perfect in every way. Remember I told you that they were not that much different than us. We grew very close during that time and we both fell in love. She loved me and showed me more love than I had ever experienced in my life, so we had a baby together." "What, Josh says? A baby?" "Yes. We named her Angelia." "Like momma," Sammy says. Their eyes widened. "Your Mother," Dr. Mitchel says. The kids' mouths fly open. She was the most beautiful baby I had ever seen. "I knew it! I always knew mom was an alien," Josh says. "That explains a lot." "That's why she's so tall too. I think so," Dr. Mitchel says. "That means we have Adimowan blood too," Josh says. "Yes, you always have. We were a little family for a while, but it was short-lived.

Makia wanted nothing more than to stay here with us, but her father, Makal, came down and took her home I guess. He forced her to go back. They knew there was a baby involved but it didn't matter. She stayed behind with me. I raised her alone. When your mother had questions I told her that her mother died giving birth to her. I had to lie about it. Your great-grandmother Anne, God rest her soul, helped me to raise her. At that time I had no idea how to raise a child, especially a little girl. She was raised just like any normal child, but her special gifts could be seen at times. She was always at the very top of her class and was an exceptional athlete in most sports. With the help of your great-grandma, I was allowed to continue my life's work, but eventually, the projects ended and I was done. I was a single parent now. I haven't seen Makia in like thirty years. There's not a day that goes by that I don't think of her though. I just hope that she's doing ok up there. She told me exactly how to get to Adimowa without dying, but honestly, I've always been afraid to try. I stayed for your mother's sake, I just couldn't take that chance that something would go wrong and she'd be down here all alone." "So, what's in your book?" asks Mona? "Everything we need to know and then some. The kids' eyes widened at the thought of that, of everything you need to know and more. "Everything?" Josh asked. "Yes, okay not everything, but I do know things that no one else in the world currently knows.

In this book, there are step-by-step directions on exactly how to get to Adimowa and back to Earth again. They all stare at the book in his lifted-up hand. Eventually, I would have gone one way or another, I guess. We will go to Adimowa and ask for help. Maybe that's my purpose. I will go there and talk to Makal the Arc. "Your father-in-law, right," Josh asks. "Yeah, I guess so." Oh, he probably won't be happy to see you at all. It was a long time ago, Dr. Mitchel says, hoping that all is forgiven by now. "Maybe you can see her when we get there," Mona says mischievously. Dr. Mitchel paces back and forth while rubbing his grey head. I can't think about that right now, he says. He opens his journal to a particular page and reads it. There are only three ways to get Adimowa, we can fly, or die or translate there. Flying there would take too long, we would die on the way. Ironically, we'd be in Adimowa but then we couldn't come back because we'd be dead. They all look at him with a very confused look on their faces as he explains. We need to translate there, it's the only way. Makia told me that there's only one point in this world where you can do it. "Where?" Sammy asks. Dr. Mitchel shows them a page from the book. "Point zero," Dr. Mitchel says with excitement. "If we can get to point zero, introduce the right sequence of special stones and incantations, we will be teleported or translated to Adimowa. According to this, we would come out of the River of Life in Adimowa.

They just look at him, not knowing what to say to him. At that moment Sam Cooke starts to sing from Dr. Mitchel's old-style radio " I was born by the River". No way, they all say, "that's got to be a sign Pawpaw" Mona says. I think so, he says. When going that way, we'll have exactly twenty-four Earth hours to get back into the river and translate back to this realm. If we don't make it back, we will be stuck there forever. We will go to the center of

the   world."   "Where   is
that?" Josh asked. "Let me show you." He pulls
out his journal and shows the kids his notes
that demonstrate how Egypt is at the
centermost portion of the world's landmass.
We will go the way of the pharaohs, through
the King's Chamber. The Great Pyramid in
Cairo is point zero. "Let me show you
something. You see, it's the place where
everything began, the place where past,
present, and future all come together.
Remember the beam of light they have
coming out of the top of the pyramid in Las
Vegas? "Oh yeah," Josh says. The truth is
always hidden in plain sight, Dr. Mitchel says.
"Oh yea," says Josh. "Yup. Einstein had the
right idea all along with his theory of relativity.
The idea of black holes that could take you
through time and space. Think of the Great
Pyramid of Cheops as a giant prism reflecting
light and time in every direction.

We just need to patch into the direction that
goes to Adimowa, and I know the way. It was

also known as Jacob's Ladder. Right after he says that they all hear something. It's like a cross between an angry rabbit wolf and a rhinoceros rumbling towards them from inside the cave facility. "What was that?" asks Mona. "Get your stuff we're leaving right now." They quickly grab their packs. "Whatever it is, it sounds big." The sound gets louder and closer still. "They must have left something. That's bad." "What do you mean...*something?*" Josh asks. "I think it might be somebody's mutated science project left behind." "It's right there!" Sammy yells. Mona screams. "Go!" Josh yells. With everything they've got, they run to the ship. "I left my book!" he yells. Dr. Mitchel runs back to the table to grab it, while the kids keep running for their lives. They barely make it into the ship when the beast arrives. The kids see it from the ship's bay window. "What is that?" Mona yells. "Hurry Papaw!" "That's not from this planet," Sammy says. Dr. Mitchel keeps running as best he can for an old man. He is completely exhausted, but finally makes it into the ship. "Close the door!" he yells. Josh hits the button that seals the door.

With a face only a mother could love, the hideous beast leaps onto the Issachar, rocking it. They struggle to stand, then fall to the floor. "Are you ok?" says Dr. Mitchel. They all say yes. "Let's get out of here!" From the outside of the cave, you can hear and feel its roar. Snow and ice crack and fall at the mouth of the cave from the vibrations. "We got to go!" Dr. Mitchel points at the mouth of the cave, the ice has started to block the door. The beast continues to gnaw and rip at the Issachar's outer hull. "He's trying to get in here," Sammy yells. "He can't do that." Dr. Mitchel explains that the Issachar is virtually indestructible. He fires up the Issachar that has been on flight standby since they arrived. Even starting the ship doesn't remove or disturb the creature in any way. His enormous, tentacle-like arms have a firm grip. The ship vertically takes off, then turns towards the mouth of the cave. It starts to increase speed heading to the door that is now almost completely blocked by thick ice. "He won't let go!" Mona, screams. He will. The ship begins to pick up speed as it gets closer and closer to the exit. The opening is almost totally sealed, with just enough room for the ship to barely pass through. "We're not going to make it!" Josh yells. "Yes, we are, I got this," Dr. Mitchel says. They barely make it through what's left of the opening as the door completely seals behind them. "That was very close. Too close," Sammy says.

"Where to now?" Josh asks. "We're going to Egypt to translate, but first, we need to go to Florida. We got to get to your mom and dad; they must go with us. "He's not my dad," Josh says, in an angry voice. Sammy and Mona look at Josh with a look of disbelief. Dr. Mitchel knows right away that he must deal with what he just said very carefully. He puts the ship on autopilot in an orbiting holding pattern. He takes his seatbelt off and turns towards Josh. "Hey Josh, I know that things haven't been easy for you and you have experienced way more negative stuff than your brother and sister. Louis is their father and the only father that you've ever known. You have to respect the fact that he raised you too. I know that they fight like cats and dogs but they do love each other and they definitely love you guys. That may be the one thing they both agree on. Just try to be sensitive for their sake. He points at Sammy and Mona. Josh starts to cry and nods yes. "I'm sorry," he says, "It's just hard for me sometimes." "We know," says Mona. "We're family," Sammy says. "Now let's go get your parents."

Dr. Mitchel and the kids begin to freely move about the cabin. "We've got to let your mom and dad know what's going on. I know that they are having problems right now, but this is bigger than that. If this doesn't bring them together, I don't know what will. I know, Josh says. It will be alright, you'll see. No matter what happens we're family. You want to know something; I know that they do love each other, and they will stay together.

"How do you know that, how can you say that Pawpaw?" asks Mona. "I don't know. Maybe it's the Adimowan side of me, but I know it. A real change is coming." "I hope so," she replies. "Me too," Sammy says. "What about you Josh?" He smiles and says, "I'm not going to hold my breath." I'm older than you and have way more fond memories than both of you. I hope Mom and Louis, I mean Mom and Dad change. "I hear you, Josh," Dr. Mitchel says. "Wow, this is some heavy stuff we're talking about," Dr. Mitchel says. "But this is good, you need to talk about it." They all nod yes. "Ok, enough of that, let's think about what we need to do."

Camp David Presidential Retreat (Shangri-La)

Catoctin Mountain Park, Maryland

July 14, 2021, 12:07 AM (ET)

The President is wide awake while his family all rest quietly at Camp David retreat ranch in Frederick County Maryland. Curtis, his personal butler at Camp David, brings him a hot tea right on time. "Knock, knock," he says cautiously. I thought you might need this Sir," he says. "Perfect Curtis, thank you so much," The President says. "Will you be needing anything else Mr. President?", he kindly asks. "No thank you, the President says. Good night Curtis." "Good night Sir," Curtis says as he closes the office door. He contemplates the future of his family, the country, and Callisto. U.S. Marines stand a watchful guard at every quadrant around the perimeter of the residence. The President thinks of all the past presidents and their families and staff that passed through these halls and slept safely in these rooms. So much American history here. "If these walls could talk," he thinks, as he sits at his desk. He reaches inside the right side desktop drawer and pulls out his presidential letterhead paper and pen. "Well here we go Koda," he says to his faithful German Shepherd that sits at his feet. He begins to think of every person in his life right now, family, friends, business associates, enemies. He gets his opportunity to be a god, just like Pherous said he would. Without any emotional moments or really any feeling at all, he creates the list of those who will go with him to Callisto and those who definitely will not be going. It only takes him a few hours to form his "knotty or nice" list. A soft knock at the door then it slowly opens. It's the President's wife Gladys. "Why are you up so late Randolph, we came here to get some rest and we're leaving tomorrow for Switzerland. You promised that you would not do any work." I know dear, but this couldn't wait, I'm sorry," he says. She walks over to him seated at his desk and hugs his head.

"Well at least you had some tea to keep you warm, she says. What are you working on that's so important that it made you break a promise to me." He slowly looks up at her. "Oh my God, look at your eyes, they're so red," she says. "I had to make lists," he says. "What lists," she asks. "Come here," he says. She does. "Yes,

she says, I know that look, what did you do,"
she asks. "Life is going to change for the entire
world very soon Gladys. "What are you talking
about Randolph?" she asks. "I'm talking about
the end of the world as we know it, but also a
new beginning. "You're serious," she asks. "Yes I
am, he says back. We made contact with the
beings that created us and they have told me
that the world will be completely destroyed.
But we have a plan, Gladys. We will be saved,
they're going to make a new planet for the
"Chosen." Suddenly she understands what
these lists are he mentioned. Our family will
be safe and we will have a future. She quietly
listens to him. As President of the United
States, I was sworn to secrecy and couldn't
even tell you about these beings or aliens, but
they have returned to Earth. When they come
back It's called a visitation, this is the last one.
They have been coming to Earth and going
back to Admiowa for centuries. "Adimowa," she
asks. That's the name of the place they come
from. It might as well be called Heaven. No

President since Eisenhauer has had to deal with this. I was chosen to receive them this time because it's the end of Earth. "I have to decide who will go and who will stay here and die on Earth when it is destroyed. "What!, she says. Oh, you love this don't you, you get to play God. You finally get to be god. So who gets to go?", she asks. He corrects her. "I finally get to be "a god," he says. I guess you better start showing me more respect then, he says facetiously. "Good night Randolph, she says angrily as she walks out. "Good night babe." I am a god, he says with a devious look on his face.

CHAPTER 7

OUR GENESIS

GENEVA SWITZERLAND, JULY 14, 2021 6:18 PM (CET)

The Hotel du Parc Eaux-Vives, is completely booked and blocked off for nothing but the hand-selected very rich, educated, and wealthy families of the world. The kind of rich you'll never meet. The extremely, very well-dressed crowd of very special people anxiously waits for the speaker to take the podium in the crowded conference room. You can feel a tangible buzz of excitement in the room. Chairman Jason Roth arrives and walks to the platform. The crowd goes to a silent hush as all eyes fixate on him. "It gives me great pleasure to say welcome to the New World Council of Callisto! Gentlemen, ladies, distinguished men, and women. You each represent the chosen. The future of mankind is right here, right here in this room. The dawning of a new age has come. You have all been carefully vetted and selected to be the families that cross over into this "New World Order." This moment signifies the height of our achievements.

This is our Genesis moment, our beginning. The terraforming process of Callisto, the fourth moon of Jupiter, has almost been completed and will be ready for us very soon." The crowd claps wildly. "Pherous, we thank you; we thank you for this precious gift you have given us. "All hail Lord Pherous!" They all say it back to the chairman in unison. "All hail Lord Pherous!" Hundreds of adoring Pherous worshipers lay at the feet of his 12-foot black onyx statue. They quietly sob and lay flowers and fruit down. They blindly worship Pherous even though the vast majority have never even seen him. They don't even question why the statue is black onyx stone. "Ladies and gentlemen, the

President of the United States of Callisto," Chairman Roth says. The crowd claps for several minutes before slowing down and then stops. "Thank you, Mr. Chairman. All hail Pherous! All hail Pherous!" the President shouts. All together they repeat what he said. "Thank you so much. Please take your seats and please hold your applause until the end. First, I want to formally thank the G7 and the chosen ones for choosing me to be your representative and allowing me to carry this torch for humanity. What was once our precious Earth is certainly in trouble today.

I need not tell you how it is dying, even now, right before our eyes. I know it pains us all to admit this. Despite the global pandemics we experienced over the past few years, the population continues to grow to unsustainable levels. I need not remind you of this fact. The ozone layer is depleted, acid rain, mercury in the fish, killer bees, global warming, the list goes on and on. But the greatest and most looming threat to our existence is the threat of complete extermination at the hands of our creators. We cannot allow this; we shall not allow this. Space Force was created for this reason, but it was never going to be enough to protect us completely. Pherous has promised us and more importantly delivered on his promises. Life and not death! Sustainable life on our new Earth. Long live Callisto! If our creator plans to wipe us from the face of his Earth, it will not be without a fight!" The people roar with excitement. "We act now to preserve humanity and our right to live. The time to move on project Callisto has come and move we shall! He pauses to receive the crowd's praises. We are going to destroy this miserable planet with the help of our dear friend Pherous. We are creating a better place for mankind, for those who deserve it.

We will not make the same mistakes that we made in the past. It will be a new beginning. It is time. We are the generation to do this, and we will complete this monumental task. Callisto is waiting for us! The fundamental question of who will go and who will stay must also be answered now for the benefit of all mankind. It has been agreed that sacrifices must be made in order to form a more perfect new world union in Callisto," the President said. "We will start over, leaving behind the things that caused our global demise. The poor, the weak and uneducated fools and criminals who kill the innocent without regard. Those who seek to make war through acts of barbarism and terrorism. Those who cannot see and will never see the way. The idiots who believe that voting changes things. They say I'm going to write my congressmen. I'm going to give them a piece of my mind. Well, you go right ahead and do that. All those religious fanatics who profit from the fears that they create and exploit the ignorant masses. Our inept politics have crippled us for far too long, they must be left behind.

Allow me to paint the picture. Somewhere in a ghetto a child laughs and plays on a trash-filled playground. Gunshots ring out. A single mother on welfare cries for her son that lies bleeding to death on the street, the innocent victim of a drive-by shooting. In an alley lies a homeless man on a piece of urine-stained cardboard, yelling obscenities at a streetlight just before he beats an old woman to death. Can you see it? A cop shoots an unarmed black man in the back while his hands are raised in the air. Can you hear the chants of "no justice no peace" and "Black Lives Matter" filling the air? Riots and looting have now come to the street you live on. But wait there's more. Somewhere in the world, a terrorist makes their insane demands, while threatening to cut off your son's head on national television. Street gang members fight other gang members for corners that they don't even own. I won't lose any sleep over that. The crowd chuckles. We will find true freedom and equality, but we must shed the unnecessary weight that has been allowed to drag us into ruin. We must take courage and show the resilience that will make us great again. Sleep well tonight. On to Callisto!" The room goes crazy with excitement, cheering, "Callisto matters! Callisto matters! Callisto matters!" making a complete mockery. Revealing this really is, a shame before God. Thank you. The people cheer as the President takes his seat. Chairman Roth approaches the podium. "Thank you, Mister President. All hail Lord Pherous! All hail Lord Pherous," they all shout.

Callisto (the New Earth)
July 14, 2021, 9:27 PM (Space-Time)

Water starts to spring up from the ground and flows into new rivers and oceans. The surface of Callisto starts to rumble and quake. Volcanoes erupt all over and mountains sprout up into the air thousands of feet. Clouds are formed. Hurricane-force winds start to blow. "Start the core heating process now", the engineer bellows. More rumbles and grumbles from the guts of Callisto. Steam shoots up from the cracks in the ground, while a storm continues to rage and swirl in the newly created sky. Monsoonal rains saturate the ground. At a nearby dock, animals of all kinds begin to arrive and are quickly taken to the animal storage facility. Green trees and plants grow up instantaneously all over. The team of engineers and scientists work the machinery and turbines above and below the surface of Callisto. "It's working, we have done it!" the second engineer yells. The engineers and scientists randomly hug each other and jump for joy.

They run outside and look up as the rain falls into their faces. The air is pure and perfect. The staff of scientists and engineers all lift their hands in silent reverence to Pherous. They think that it is more beautiful than the Garden of Eden ever could've been. The process of terraforming Callisto is finally complete. Even Callisto's moon and small man-made sun are in place and orbiting, creating only a 12 hour day cycle. Callisto is just like Earth in every way except one, it has yet to be defiled by men. "Ok, let's get moving and start to put up the dwellings and other essential infrastructure," the main engineer says to the rest of the crew. They immediately start to move out into the rest of the capital area of Callisto and begin to build. Pherous hovers high overhead and marvels at his achievement. "My world is complete!" he says. "Let there be life," words that he desired to say for centuries.

East Africa, Mount Kilimanjaro

July 15, 2021, 3:36 PM (EAT)

Pherous projects himself and the President to high atop Mount Kilimanjaro in Africa for a little fireside chat. Instantly they both step through a rip in time onto the plateau. "It's so cold," the President says. Did it have to be here Pherous," he asks?" The cold doesn't seem to be a problem for Pherous at all. "Are you cold?" he asks. It feels fine to me. "You failed me," Pherous says. The President shivers and tries to warm himself by blowing hot air into his hands. "Yes, but I sent the best of the best; they almost had him, but he got away. My people told me he's like a wizard, a shaman, or something like that." "I told you not to underestimate him," Pherous says. Don't worry, we will get him, I promise you," the President says with confidence. "I'm starting to have my doubts about you Randolph," Pherous says. I don't like failure. Do you realize that I'm about to give you the entire new world?" Pherous says. "Yes, yes I do," the President stutters from being extremely cold at this point. "I have multiple teams looking for him right now. Even Interpol has him on a most-wanted list, which means that the whole world is looking for him. He won't be able to hide anywhere in the civilized world." "He is the greatest threat to what we are trying to accomplish,"Pherous says. "He has a daughter living in Florida," the President says. "A daughter, Pherous says with surprise. "Yes. He most likely will try to contact her at some point. When he does, we'll get him; he will let his guard down and we will have him," the President says. "So it's true? There is a child that lives, a daughter. Pherous' wheels begin to turn in his mind. "Yes, find her and you will find him I'm sure of it," Pherous says. "That's literally what I just said," the President sarcastically says. Pherous gives him a look. "I'm sorry, Sir," the President says, lowering his head from eye contact with Pherous. It's still very hard for the President to grasp that he is no longer in charge or control and must show the utmost respect to Pherous. "Let's get you to somewhere a little warmer, shall we." Yes please, the President says.

CHAPTER 8
WE'RE HEADING TO FLORIDA?
WASHINGTON, D.C., JULY 14, 2021 6:56 PM (ET)

Alfonso sits at their favorite table for two by the window, in the almost empty local coffee house. Zoe waits by the bar for the barista to finish up their special orders. Alfonso glances out of the window across the street and sees a small group of blacks and Hispanic men, all with beards and wearing what looks like strange blue or purple clothing. Alfonso wonders what they are saying that seems to have the people's complete attention. He gets up and walks over to Zoe. Hey, I'll be right back, he says to Zoe. Where you going, she asks. I want to see what they are saying, he says, as he point's across the street. "Oh, them," she says somewhat condescendingly, apparently having her own opinions of the group. "Ok, but hurry back, the coffee is almost ready." "I will," he says.

Alfonso runs out of the coffee shop and crosses the street. He walks up to the crowd and sees the men holding up signs and reading from a Kings James Version of the Bible. Thanks for coming over brother, one of them says in Alfonso's direction. Alfonso just looks wide-eyed and smiles back, feeling way out of his comfort zone. "That's exactly who we are according to Deuteronomy 28:15 through 68, and guess what, we are living in the last of the last days and we don't have long. It's time, high time for you to repent as Israelites and keep God's laws, commandments, and statutes before the chariots come back to get us. This place, "Babylon The Great", will be destroyed by fire," one of them says with authority. "That's right!" they all shout. Alfonso cautiously raises his hand high to ask a genuine question. "Go ahead my brother, ask your question, the leader says. What's a chariot? "That's a very good question," he says. "The chariots are the Angels of the Lord, that's how they get around and do the assignments of the Highest." "They travel in the clouds." "They are all of these so-called UFOs that people see all the time, all over the world." "Excellent question my brother, excellent." Alfonso smiles and then fades back into the crowd and goes back across the street. He can still hear them preaching in the distance as he makes his way back across the street to the coffee shop.

Zoe sits at the table in the big window, waiting for Alfonso to return. He comes in and sits down with her. She hands him his coffee. "Well, crazy huh?", Zoe says "Wow," Alfonso says, making his eyes wide. "At least I know what a chariot is now," Alfonso says. "A what, Zoe asks? "Never mind, I can say that it was very interesting and they believe what they are saying,", Alfonso says. "They believe that God and the Angels are coming to destroy the Earth in these things called Chariots in the clouds," he says. "On a different note, it's quiet on Capitol Hill, Wall Street, hell everywhere, it's weird," Alfonso says "I can feel it too, Zoe says. "Zoe, you think the world is coming to an end, you think it will be destroyed one day like they say? Alfonso asks. "Wow, what a loaded question," Zoe says. "This planet is in trouble for sure," she says. It's dying, we are killing it, Zoe says." "I believe we already killed it and we're just seeing its slow and painful death," Alfonso says.

That's actually what those dudes across the street were saying too. "Yeah, and they're probably right," Zoe honestly replies. "By the way, I forgot to tell you what I found out through a few sources of my own. I know where Dr. Mitchel's daughter is," Alfonso says. "What! Why didn't you tell me?" Zoe says "I am telling you now," he replies. "She's in Florida, the Everglades. She's married with three kids now, two boys, one girl. "They own and operate a small airboat touring company." "Outstanding Alfonso! I guess we're going to

Florida," Zoe says. "Do you think they have gators down there?" Alfonso just smiles. "Maybe she knows where her father is. "Do you have an address for me?" "Right here," he says. Alfonso hands over a small scrap of paper to Zoe. "You should be a reporter sir." "I'll get the bill, you've earned it," Zoe says.

Florida Everglades Airboat Rides
July 14, 2021, 3:53 PM (ET)

Angelia and her husband Louis are having another heated argument in the lobby of their Fish and Airboat tour business. Business has not been so good lately due to the several hurricanes they have experienced. The sound of breaking glass causes Louis to flinch. "You almost hit me in the head!" exclaims Louis. "You better be careful." "Look, it's "hurricane Angelia." Ha-ha. You make me sick to my stomach! Do you always have to act this way?" "You're a complete idiot sometimes, you know that," Louis says. "Yeah, I know it, but you still love me," Angelia says. A customer walks in and hears the fight, then turns and walks right back out the door. "No, come back," Louis says. "Damn. You see, that's why nobody comes in here," Louis says. "This marriage is a bad joke," Angelia says. "You want out? There's the door. Maybe you can go live with your daddy." "Maybe I will and go get my groove back," Angelia says. "And leave our little slice of Heaven behind?" Louis sarcastically asks. The small bell over the front door chimes as the door opens. Zoe and Alfonso cautiously walk through the front door hearing the tail end of the argument. Angelia and Louis notice them coming and try to act as if nothing was just happening

"How can we help you folks today?" Louis asks. "No, we're not customers." "Are you bill collectors or realtors; if so, we aren't buying? "No. I'm Zoe Baker and this is Alfonzo Nazario, my assistant. Alfonso smiles and says "nice to meet you." Is Dr. Horace Mitchel your father, Zoe asks?" directing her question towards Angelia. "Maybe, why you want to know?" "I'm a journalist with KNRA news." "Never heard of it," Louis says. "No one has," Zoe replies with a smile. That kind of breaks the ice between them. She shows them her press identification. "Ok. So what do you want?" " Do you know where your father is?" "Why are you looking for my father? What did he do now?" "We think that he may be in danger." Louis goes over to the front door and turns the open sign to closed. "You have my attention now Zoe. Why is he in danger?" Angelia asks. "I'll get right to it. Was your father involved in any secret projects for the U.S. Government?"

"My dad did a lot of things in his life that I know nothing about. I know that he was some kind of special type of Marine and he worked for NASA. He never said anything about any secret projects to me." "Can you tell me where your father is? Why is he in danger? "Your father was involved and worked on many top-secret projects for the U.S. Government. One of those projects is getting people killed and or coming up missing. We need to find him and crack this wide open. If we break the story, they will have to back off. Wherever he is they are probably going to try to get him." Angelia thinks about the fact that her kids are with him right now. "I also believe that there is something about to happen on a global scale. I think that your father might be involved in some way or has something they want." "They who," Angelia asks? It's complicated Angelia but we need to know where he is, Alfonso says. "He's in Jamaica, that's where he lives now. He retired there years ago. Angelia looks at Louis with a terrified look on her face. "Our kids are there with him for the Summer. They go for two weeks every year." "We really need to speak with him, Angelia. "Sounds like I do too," Angelia says. "Can you please call them?" Zoe asks. Angelia runs to grab the phone while Louis just stands there in shock. She calls her father's house several times, but no one answers. "We need to go there right now!" Angelia starts to get worried.

While Zoe speaks to Angelia, three men in black suits look through the window from the outside. Louis changed the sign to closed but didn't lock the door. One of the men sees them and tries the door. The tiny hanging bell above the door rings as they enter. "Can we help you gentlemen Angelia asks?." "Did your limo break down or something," Louis jokingly asks. "That's funny big boy," one of them says, smiling and looking at the other men. Alfonso repositions himself and almost takes a fighting-ready stance, sensing that this is not going to go well. Zoe and Angelia take a step back. "Yes, I think you can help me with something." The man violently grabs Louis by his throat, with both hands. Instantly chaos breaks out in the shop. "Let him go!" Angelia yells. Alfonso tries to step in and help but is quickly punched in the face by the biggest one. He goes down hard. "Damn", one of the men says. Zoe goes to the ground to help him up. "Alfonso! You didn't need to do that you asshole!" "Oh, I think I really did," the man says. Jeff pulls out a

45-caliber handgun from under his jacket, while the others kick Louis while he's down. Louis is knocked unconscious for a few seconds. "Are you Angelia?" Jeff asks. "Yes," she says sobbing. He speaks into his wrist communication device; we have her Sir. Pherous receives the good news. "Excellent," he says. He will be there soon, his energy is on the move. "You know what to do. Have the chains there. They must be old and rusted, do you understand?" Yes Sir," Jeff says, rusted chains. When that's done, we will be ready to move forward. I will be there shortly. I want to speak to him face to face. He wants nothing more than to be there when they capture Horace and his Adimowan daughter. Pherous uses the power of his mind to conjure up his transportation. He moves into the fiery cloud and bolts across the sky, heading to Florida.

Dr. Mitchel and the kids arrive in Florida. As they descend, he decides to land the craft in the nearby marshland of the Everglades. This spot is not far from his daughter and son-in-law's airboat rental and bait store. The ship is placed on shroud mode, which makes it completely invisible to the naked eye. Dr. Mitchel lands the craft and again puts it in standby mode. He gently wakes the kids from their long naps. The kids begin to stir and wipe their eyes as they awaken from a nap, some much-needed sleep. They take off their Adimowan power napping devices and start to unbuckle their seat belts. "We're in Florida?" Josh asks. "Yep, we made it," Dr. Mitchel says. "Momma!" Sammy yells. "This area is perfect, no one will be able to see it. Someone would have to bump into it to know it was even here. Let's go, we'll be able to catch them both at the shop." They gather a few essentials into their backpacks and head out. The boys start to head toward the exit. Dr. Mitchel pulls Mona to the side. "Mona, I have a special mission just for you." "What is it?" Mona asks. He takes out his journal and hands it to her. "I need you to hold on to this for me just in case." "Just in case of what?" "In case they catch us", he says. Mona is happy because he trusts her enough to give her his special book. "I won't let anything happen to it; I promise." "I know, that's why I gave it to you," he says. She hides it on her body, tying it to her waist at the small of her back. "Ok, Let's go," Dr. Mitchel says. "It's strange walking down invisible stairs," Mona says. "Yeah," Sammy says. "Try to hold onto the rails," Dr. Mitchel says. As they get down, they look toward the tail end of the ship and make a startling discovery. "Holy crap!" Dr. Mitchel yells. "What in the world?" Mona says.

The creature from the Antarctic cave is still latched onto the ship but clearly dead. "It's still there," Josh says. It looks as if it is floating in midair as it rests on the invisible tail of the ship. "It's dead, I think," Dr. Mitchel says. "Just in case let's get out of here," Sammy says. They all agree and start running away. "Stay close everybody, they're alligators around here," Dr. Mitchel warns. "Be careful where you step."

They carefully walk through the marshlands. After a few minutes of travel, they make it safely to the end of the marsh and stop. "There it is," Dr. Mitchel says. They hide behind a group of trees just across the street from the shop. He takes out a special binocular device and surveys the shop. "Everything looks normal," Dr. Mitchel says. "Let's go." They cautiously approach the front door. The kids get excited and run through the door. The door chimes above ring out. They see their mother and father standing behind the counter like nothing ever happened. "Momma!" Sammy yells. Zoe and Alfonso are also standing there, pretending to be customers and looking to book an airboat tour. Dr. Mitchel feels that something's not right. "Hey kids," Louis says. He looks from side to side, with a scared look on his swollen face. "What happened to your face?" Mona asks. The tension can be felt in the air. "Angie, is everything ok?" Dr. Mitchel asks. She shakes her head and half-smiles. Dr. Mitchel and the kids assume that they must have had yet another bad fight or something like that.

At that moment two men in black suits walk into the shop. Ring goes the bells hanging over the door. Team Leader Jeff Aragon comes walking out from the back of the shop with his gun drawn. "This guy again," Dr. Mitchel says. Jeff puts his gun back into his shoulder holster. He takes out a cigar and lights it up. He blows out a large puff of smoke into the air, then says, "Don't even try to run, you won't get very far, old man." He walks over to Dr. Mitchel. "I owe you this one Raider. Jeff grits his teeth and musters up all the energy he can find, then sucker punches Dr. Mitchel right in his face. His glasses fly off his face. He falls to the floor like a sack of potatoes, knocking over an endcap display of fishing rods. His nose erupts with blood all over the front of his shirt, and his lip is split wide-open. Dr. Mitchel is still conscious but is in a world of pain. He spits blood out onto the shop floor and then looks at Jeff. "What the fuck was that for!" Angelia shouts. "Why did you do that? she asks. The others have to be held back by the other men. It's a scene of complete chaos. "Damn, he can still take a punch," a team member says begrudgingly. He gives one of the other men in black a twenty-dollar bill. He won the bet. So much for "You knock me out, I knock you out, Jeff." "Get them out of here," Jeff says. He rubs his fist and shakes out the pain. Josh and Sammy help Dr. Mitchel to stand up. The men put glowing red, futuristic-looking handcuffs on each of their wrists.

"Do you have to put those on the kids?" Louis asks. The man nods affirmatively. "Real tough guys," Zoe says sarcastically. Mona starts to cry. Josh puts his arms around Mona. "Don't worry sweetie, I won't hurt you," Jeff says to Mona. Mona rolls her eyes at him. "Don't even try to get out of them; it's impossible," Jeff says. Dr. Mitchel recognizes the technology as Adimowan in origin. Now he knows they are getting help from a visitor. "Dad, what the hell is going on?" Angelia asks. "The end of days baby," he says. "Shut up!" Jeff says. Ok, let's get them loaded up, he should be here very soon. Search them, first. The men and women get searched from head to toe but the men find nothing except Dr. Mitchel's heart medication. "Dang, you all messed up," one of the men says. They look at Jeff for his approval. He gestures that it's ok for him to keep the medication. They forget to search Sammy, Josh, or Mona. "Leave him here," Jeff says. He wants to see him. They load the rest of them but keep Dr. Mitchel standing there, barely able to stand. Just then Pherous and his two royal Adimowan fallen guards arrive through a rip in time, out of nowhere from a fiery cloud they appear and step out. "Horus, old friend," Pherous says. "Do I know you," Dr. Mitchel says looking up to see Pherous?" Dr. Mitchel asks. "No, but I know you.

You are the one that was chosen for the last visitation, in your year 1979. You are famous in Adimowa," Pherous laughs. The other two Admowans smile slightly but say nothing at all.

"Famous, for what? Dr. Mitchel asks. You violated a female Adimowan royal, shamed her, something that hadn't happened in thousands of years. "Oh," Dr. Mitchel says. Her father was very upset, I'm very surprised that he hasn't called you back to the river yet. "Soon enough I guess," Dr. Mitchel says. The shame that you brought to that family could not be undone. Dr. Mitchel just looks at him and thinks about Makia. "Does your daughter know yet?" Pherous asks. "No, she doesn't," Dr. Mitchel admits. "You should tell her, it's not good to keep secrets like that or lie. She is very beautiful and half Adimowan. She's special." Pherous turns to his two royal guards. "Maybe I'll make her my slave." They both nod yes. "Put him in the chains!" Jeff motions over to his men to take him. They grab Dr. Mitchel who doesn't put up any resistance. "Take him to the Place of the Great Rock.

The energy there will aid us in holding him," Pherous says. "Yes sir," Jeff says. Dr. Mitchel tries to think of where the Place of the Great

Rock might be. He remembers Makia telling him that certain locations on Earth are like antennas or conductors of energy. Places like Easter Island, Yellow Stone, and Machu Picchu. Where was this Place of the Great Rock he referred to? They drag in some old, rusted-out iron chains and wrap them around Dr. Mitchel's hands and feet, securing them with a large brass lock. Pherous raises his arms and says words under his breath, then touches the chains with his now glowing right hand. They put him with the others. They open the back of the van and toss him in headfirst, then slam the doors shut. "Are you ok?" Angelia asks. Dr. Mitchel can hardly speak. "Yes, somebody please get me my nitro pills," he says. Angelia grabs them from his front pants pocket and gives him one. He must take it dry. "Thank you," he says. " I feel strange," Dr. Mitchel says. The kids and Angelia embrace Dr. Mitchel as much as possible in the electromagnetic cuffs. "How are you going to get us out of this one Dad?" she asks. "What the fudge?" he exclaims modifying his curse words for the children's sake. "Why are you the only one in those chains? Angelia asks. "Do you know what this is about?" "Yes, please tell us," Zoe says harshly. "I do. I will," he stutters. "I'm so sorry I got you all into this mess. I had no idea that they would come for me and my family over something I did that long ago. I was working for my government. Or so I thought." "Wait a minute, who are you, looking at Zoe and Alfonso?" "That's Zoe Baker and Alfonso Nazario, they're reporters and they've told us more than you," Angelia says. "They came to get the story and I guess to help you. Who are these men?" "They are agents of the U.S. Government, Ghost Recon I think, sent to get me. Maybe silence me. I think they want something I have. Maybe my journal. The first team they sent to Jamaica said they were ordered not to hurt me but to detain me until something was over. I don't think they will kill us."

"The first team. Dad, you put my kids in danger; hell, you put us all in danger," Louis says angrily. "You're right, but I had no idea they were coming. As soon as I did, we got the hell out of there. I'm sorry, but this has been hard on me too. Look at me." They all look at the old man in chains with a bloodied face. "Ok, so can we move on now and stop playing the blame game," Alfonso says. "Why is this happening and how do we get out of it?" "It has something to do with the project I was a part of back in 79." Dr. Mitchel turns and looks at the kids sitting on the ground and sobbing. He knows that it's a long story that must be told. He turns and looks at everyone else. He takes a long deep breath. I was on the project that received what are called visitors." "Visitors," Angelia asks. "What kind of visitors?" Zoe asks. Visitors from where Alfonso asks. Ok, keep an open mind, he says. Visitors from a faraway place called Adimowa." While carefully listening to him they all almost forget that they are prisoners in the back of a moving van. Adimowa is what we think of as Heaven. It's the place where the creators live. "The creators, Heaven?" Angelia asks. The kids look at the other adults and nod yes. "Yes, they created us," Dr. Mitchel goes on to say. Remember that project that I was on in 79, it involved me living with a visitor for three years, just me and her in a bubble.

I was chosen for that visitation. You see, a lot of things can happen between people in three years together. They all look around at each other with a look of confusion. Her name was Makia a princess of Adimowa. He looks at Angelia and then pauses for a few moments. "Makia is your mother," he says looking into Angelia's eyes." Angelia's eyes roll back into her head, and then she faints. No one is behind her to catch her. "Ah damn," Louis says as he goes down to help her. Mona starts to fan her. Give her some air, Dr. Mitchel says. "Mom, the boys shout!" She quickly regains consciousness. "Baby, I'm sorry I lied to you your whole life, but how could I tell you that your mother was an alien from another planet. I couldn't tell anyone that. They would have locked me up and taken you away from me." "Is that the truth?" Angelia asks. "Yes, every bit." "What a story, I have to get this out there," Zoe says to Alfonso. "This is no story Zoe, it's a reality check," Alfonso says. "Is she still alive?" Angelia asks. "Yes," Dr. Mitchel says. They live forever unless the Highest sends them to outer darkness or if their light goes out. We came here to get you both and take you to Adimowa with us. "What?" she says. "You can take us there?" she asks. "Yes, I can. She showed me the way.

"I'm sorry but you sound insane," Louis says. "I hope you guys aren't buying this crap." "I mean no disrespect Doc. "Louis, I'm telling you the truth. We came here to get you guys because I believe the international government is planning to destroy this planet and start over on another one." A feeling of terror and dread come over them all. Suddenly, the van doors fly open, and someone tosses in a canister of gas. "Good night!" he shouts. They quickly close the doors, locking them in. "It's gas," Dr. Mitchel shouts. They try to get away from the cloud by gathering at the back of the van. The gas quickly overtakes them as it starts to work. "I love you Dad!" Angelia shouts! "I love you more," he replies. They hold the kids as tight as they can. One by one they pass out onto the floor of the van. The commotion stops inside of the van. The doors unlock and someone opens them. "They're all out Sir," says one of the men. "Get them to Spain, says Jeff."

CHAPTER 9

FIND A WAY TO ESCAPE

GENEVA SWITZERLAND JULY 15, 2021 9:45 PM (EUT)

The President and the other world leaders of the G7 have one last gathering, again at the Hotel du Parc Eaux-Vives. In less than 24 hours we will begin the colonization phase on the fourth moon of Jupiter, Callisto. "I'm proud to say that we have done it, we have accomplished our goal. We will provide a way of escape from those who would seek to destroy us. The terraforming process has been completed, and Callisto is ready to sustain human life. The crowd claps and cheers wildly. The brave men and women, the scientists and engineers that we have on Callisto, with the advanced help from Pherous. It is done! They say that they are breathing oxygen normally and that the water, plant life, and vegetation, are like that of Earth. It is clean, the air is sweet, the land of milk and honey. We will know exactly how Adam and Eve felt in the Garden of Eden."

A man comes up to the podium and whispers something into the President's bent ear. "Thank you," he says. "I've just received more wonderful news. We are ready my friends, we are truly ready. The dispatches detailing the staging area are going out to the families of Callisto as we speak. Enjoy the refreshments and have the time of your lives, for you are the chosen." Waiters and servants walk around offering the wealthy aristocrat's trays of hors d'oeuvres and glasses filled with the best champagne money can buy. The help must all wear headphones that cancel out all sound. They are basically serving everyone without any hearing. "If any of them take those off, you know what to do," the President says to his guards. "We will speak to them harshly," one of them says. "Exactly," the

President says, knowing what they really mean. At the south end of Hotel du Parc Eaux-Vives, several very rich men gather in a small circle of secret conversation. "Gentlemen, our not so little investment is about to payoff big time. Gentlemen, we are establishing the new world just like Columbus and we have been chosen to populate it. We will leave this world behind very soon." They all smile and listen to him. "We have no room for errors, your lives and the lives of your family members depend on it. You must not tell anyone about this. We would be ripped apart by those that cannot come. This is our Noah's Ark moment. We will have to seal the door and have the strength to not open it again, no matter what. We will wait until the very last possible moment before telling our families the wonderful news. We can't even trust them to not talk about it." "Where will we meet?" one of them asks.

The staging area will be relayed to me twenty- four hours before our departure; then

I will provide you with the actual rendezvous location instructions. Do you understand?" he asks. They all say a vigorous yes. "Gentlemen, we will be the new founding fathers and this time we will get it right. I am proud to announce that in exactly seventy-two hours the terraforming process on Callisto will be complete." The excited crowd claps. "We will begin the preliminary review of the conditions on Callisto and then begin the colonization process. The chosen families will be notified and the staging for departure will commence shortly thereafter. Just one week from today, we will leave and establish Callisto." The other men stand and look on in astonishment, as they give each other the secret handshake in affirmation. "We are the illuminated ones now."

Private Airplane.

Cargo area Over Lisbon,

Portugal July 16, 2021, 10:37 AM (CET)

The group is in the cargo holding area of a plane. They all start to slowly wake up. "I'm so glad that we're not dead," Louis says, in a sarcastic but thankful voice. "Yeah, but where are we?" Alfonso asks. Zoe, wake up," he says while shaking her lightly. She begins to wake up. "We're flying, I hate flying," she says in a groggy voice. "How long have we been out?" Josh asks. "I'm not sure," Alfonso says. Mona and Sammy wake along with Angelia and Louis. "Dad!" shouts Angelia, thinking Dr. Mitchel died in his sleep. "I'm awake, just resting my eyes," he says. She looks relieved. Dr. Mitchel looks down to check his watch but it's been removed from his wrist. "They took our watches. I wonder why?" Alfonso responds. "That's obvious; they want to totally disorient us," Zoe says. "Well, they succeeded at that," Dr. Mitchel says. "They took everything except my medication." He looks over at Mona. She smiles. "I'm so sore, everything hurts," he continues.

Suddenly they hear the flight gear drop down in preparation for landing. They feel the plane slowly descending from the sky. "It sounds like we're getting ready to land," Dr. Mitchel says. "I wonder where we are." "He said that we were being taken to the Place of the Great Rock. Where is that?" Louis asks. "I'm not sure yet," Dr. Mitchel says. "So, what's the plan, Sir?" Alfonso asks. "Wait a minute, my mother's an alien from Heaven or Adimowa, whatever. This feels like a dream. Did I dream that?" Angelia asks. Dr. Mitchel smiles lightly. "No dream," he says. "I guess we'll have to improvise as we go. We'll know more when that door opens," Dr. Mitchel says. "You think so?" Louis asks, sarcastically. Angelia gives him that look that wives give their husbands when they say something they shouldn't have. "You look just like her, you know," Dr. Mitchel says. "No, I don't know dad! I guess I'm glad to know that she's not dead, but she might as well be." "She's a freaking alien, I'm just told! How do you tell a little girl, a teenager, a young woman getting married that her mother is from a distant planet which happens to be called Adimowa?" "There was never a right time, baby. I just wanted you to live a normal life." Angelia finally begins to accept the fact that she has a mother, who's an alien from another planet. Louis is still very skeptical of the whole thing.

"We stay together no matter what," Dr. Mitchel says. "We find a way to escape from wherever they are taking us and then we go to Egypt." Dr. Mitchel looks like hell at this point. "This is a crazy big story, I will get a Pulitzer for this one," Zoe says in an excited voice. Alfonso just looks at Zoe, amazed at how she still chooses to think everything can be normal again. "Yeah, but if what he says is true, the Daily News won't matter anymore, neither will your awards. This changes everything. If Heaven or Adimowa is a real place, and things are lining up with the bible, this should blow your mind," Alfonso says. "You're right." She starts to think about her family and friends that will be left behind. Suddenly the story doesn't seem important at all. "Before we got to Florida, I called my friend Birhan Youssef," Dr. Mitchel says. "He's the Minister of Antiquities in Cairo, Egypt. I carefully explained to him what we needed and he understood way more than I thought he would. He was a believer right away. He offered to help us but only on one condition." "What's that?" Zoe asks. He, his wife Fatimah, and two little girls, Dellina age nine, and Gabriel age fifteen, get to come with us." "Come with us where?" Louis asks. "To Adimowa! We are going to Heaven Louis."

"Dad, you know that I respect you a lot and I know that you're a very intelligent man, but I honestly think that you've been out there in the jungle smoking the herb too long. Who's with me on that one?" Louis turns to the group hoping for someone to join in. "Louis, the planet Earth will be destroyed, all gone in an instant. It's literally up to us to save it if we can or at least live to tell about it. Nothing else is more important than this," Dr. Mitchel says. They all sit down and brace themselves. The plane makes its landing and begins to slowly taxi the runway. It finally comes to a stop and then the engine powers down. It is quiet for about five minutes, then men's voices can be heard from the outside of the plane. They all stare at the cargo bay door, wondering where they are and what will happen to them next. The hatch door violently flies open. The bright light of the noonday sun blinds them. "Let's go, everybody out," says one of the rather large men.

A black van waits for them on the tarmac. "Ok, get them to the Gibraltar location and get them secure," he says in Spanish. They hurry them into the van and whisk them away. "They must be taking us to a safe house or someplace like that. Are they going to kill us?" Angelia asks. "I don't think so. If they wanted to kill us, they would have done it already," Dr. Mitchel says. "But where are they taking us?" Sammy asks. "I don't know. Just try to relax and stay calm," Dr. Mitchel says. I could've sworn that I heard them speaking in Spanish." "Well that really narrows it down," Angelia says. "Yeah, but it sounded like very proper Spanish, as the Castilian kind. I think we're in Spain. The place of the great rock is Gibraltar; we are in Spain." The van has no windows, but they all try to gain intel by listening to the sounds they hear. "We are for sure in the city," Zoe says. "They must have a police escort because they haven't stopped for any lights yet and they're really hauling ass," Louis says. "We are in Europe, you hear that siren?" Dr. Mitchel asks.

Waldorf Astoria, New York, NY
July 16, 2021, 8:25 PM (ET)

The President meets with the American chapter heads of the chosen families of Callisto one last time before they leave. "Thank you all for gathering so quickly and orderly. You demonstrate the reason why you have been chosen. The time has finally come my dear brothers and sisters of the chosen order. I have been informed that we have less than twelve hours before we leave. It is time to move. Make ready your families." The group of wealthy snobs all clap. The crowd of family leaders looks to the chairman with joy and excited anticipation. "Brothers and sisters, ladies and gentlemen, in exactly twelve hours, we will go to our prepared places, under each of your family's estates. You will get in your escape pods to the rendezvous location. If you do not get there in time you will be left behind. You must take the designated tunnel at exactly the time of notification. The exact staging location will automatically be triggered in your onboard console when you type in the code word that will be given to you on that day. Once in the corridors, your pods will automatically take you to where you need to be. The guidance system will take over at that point. You will be placed in a short cryogenic type of sleep while traveling. This will be necessary due to the speed at which you will be traveling. All over the world, the select families of the order will all be in the same place and ready to leave this miserable planet. Absolute secrecy is essential as always." The crowd claps wildly.

Pherous and the President meet in a type of white house oval office room. They discuss the arrangements they have agreed upon. "So, is everything in place now?" "Yes, he's been captured alive and is being held in rusted metal chains. Gibraltar, Spain, The Place of the Great Rock, exactly as you instructed." "Good. I need you to understand this if you can; know that when you make moves on this scale, in this realm, it tends to get his attention. We will have to move quickly. "If this does get Heaven's attention, we will be ok?" the President asks. "Don't worry, I will destroy anyone that rises against us. Unlike my father, I will do it right. I will finally avenge him. I will have my new world, and you will have yours." The President wonders who Pherous' father is. He thinks of Pherous' name to himself.

Pherous, it sounds

like something else.

CHAPTER 10

THE PLACE OF THE ROCK

GIBRALTAR SPAIN, JULY 16, 2021 6:19 PM (ET)

D r. Mitchel and the others finally arrive at an unknown destination. They are all very tired and confused, thinking that they may or may not be in Spain. They are in some kind of a jail-holding cell, in almost complete darkness. Only a small night light shines near the floor. "I'm so hungry," Sammy says. "Me too," Mona and Josh say in tandem. The security guard sits outside of the cold steel holding cell reading a book at his station desk. "Hey, help us in here; hey! They all start to call out to whoever is out there listening. Just then the security guard comes into the holding area and opens the viewing window. "Alright, alright, keep it down in there, keep it down," he says in Spanish. "Hola, do you speak English? Dr. Mitchel asks. *Si*. Yes, I do.

"Where are we, where is this place?" "Now you know that I can't tell you that. Please don't ask me that," the guard says. "We have kids in here, sir," Angelia says. "I know," he replies, "I can see them right there," he says sarcastically. "They are hungry," Angelia says angrily. "I was just about to feed you guys. Listen, I don't know what you did, and I really don't care to know. It's not my business. It's my job to watch you, That's it." He walks over to the hot food cart grabbing trays of food that look pretty good. He places them through the handcuff tray. "There you go," he says. "Thank you," they all say. "Let me know if you want extra." The guard has a sensitive side. The kids and everyone else in the cell dig in and eat like they never have before. "This is good," Louis says, "I'll take extra." "Hell yeah," Alfonso replies. "Surprisingly good," Angelia says. The kids just stuff their faces and say nothing at all. They feel as though they haven't eaten in what feels like days. They all eat real good and fill themselves up.

Dr. Mitchel, still bound, with the old rusted metal chains around his hands and feet, hops back over to the cell door. He speaks between the cracks in the door jam. "Sir, please look at us, we are not criminals in here, we didn't do anything wrong. We don't belong here. This is a mistake." The guard listens and responds from his nearby desk. "You literally sound like every inmate that has ever been in here before. I'm innocent, I was framed, those were my cousin's pants, not mine," he yells. "I suggest you just relax and try not to stress out." Dr. Mitchel tries again. "Please listen to me sir, we are on a mission for God." "Oh wow, that's a good one," he yells. "Is that a Bible you're reading sir?" Dr. Mitchel asks. "Don't try that Jedi mind trick crap on me," he says. "Yes, it's a bible, so what? I read "The Good Book" to be able to stay calm in this place. Now keep it down. A mission for God, that is a real good one," he says. "Sir we don't belong here in this cage," Zoe says. Please, just listen to him.

Dr. Mitchel asks Mona to give him his journal. She does, pulling it from under her shirt, from the small of her back. Dr. Mitchel cannot hold it himself so he asks her to hold it for him. "Turn to I believe page 297 and hold it open for me," he says. She does. They all look amazed and wonder what he is doing. Cops never search kids," he says. "Thank you, Mona," he says. She smiles at him happy that she could help in some small way. She flips through the pages looking for something to help them get out of this situation that's on page 297. Dr. Mitchel reads from page 297 under his breath. As the words from the page silently hit the atmosphere, the guard comes back to the cell observation window and opens it. He places his face in the box and looks into the cell.

"Ok, just for entertainment purposes, and I'm bored out of my freaking mind, tell me about your mission for God." "Tell him," Louis says. Dr. Mitchel pauses for a moment, just now realizing that this is his purpose for being chosen by Adimowa in the first place. The security guard looks at Dr. Mitchel with that tiny spark of curiosity in his eyes, perhaps created by the words from the journal. "I like a good story," he says. "Ok, what you got? You have exactly two minutes, starting right now." Dr. Mitchel quickly tells him all about Adimowa, Heaven, visitors, and the biblical connections between it all.

He explains everything in a nutshell. When Dr. Mitchel is finished the group looks at him. "Sir, you have to let us go, we are the only hope for the world and beyond," Dr. Mitchel says. The guard starts to slowly clap his hands. "Wow, wow what a fantastic story that was, you should write a book. No, make a big movie. I almost believed you man," he says. "What is your name?" Dr. Mitchel asks. "Rodrigo Santiago Yisrael." Dr. Mitchel looks at him with wonder in his eyes. "You're a Sephardic Jew aren't you?" he asks. "My father was," he replies, with a curious look on his face. Rodrigo looks at Dr. Mitchel with a straight face. "You are too you know, you just don't know it," Rodrigo says. Dr. Mitchel has a somewhat puzzled look on his face, curious about what he will say next. I love history, if you follow it, you find all the answers you seek." "Me too," Dr. Mitchel says. The rest of the group has no idea what they are talking about.

"You and I are the same, you know", Rodrigo says. "Some stayed here, many were expelled and some went into slavery." Dr. Mitchel is impressed by Rodrigo's knowledge about the ties between the Spanish and Portuguese inquisitions and expulsions of 1492 and the Transatlantic Slave Trade. No one teaches that history today, but some know the truth," he says. "We are the same, *Lo Mismo*," Dr. Mitchel says, finally making a connection with Rodrigo. "In the Torah, there are many examples of the Israelites encountering angelic beings from heaven, messengers, no?

Stories about clouds of fire and chariots that came down to Earth?" "Yes, I remember stories like that in there. "You are in that kind of story right now."

Rodrigo's face gets serious, and his smile goes away. He starts to think about all the biblical stories he had ever learned as a child. Stories he didn't always believe, even doubted at times. Dr. Mitchel now has his complete attention. He looks at Dr. Mitchel like he is an Angel sent from God. "Rodrigo, the Highest wants you to get us out of here," Dr. Mitchel says. Rodrigo takes a deep breath and swallows, knowing that if he helps them, he will be killed. "I can't do that," Rodrigo says, realizing what could happen to him. "Yes, you can, and you will", Dr. Mitchel says with confidence. Dr. Mitchel has been whispering words from the book in between talking to him. Rodrigo's eyes lock with Dr. Mitchel's. Like a light switch being flipped on, he takes his Adam Folger keys and opens the cell door. He swings the heavily reinforced metal door wide open for them. They all look stunned at what just happened.

"It worked," Louis says to Angelia. Dr. Mitchel is barely able to hop and the others make their way over to Rodrigo and put their hands on his shoulders. They all hug him and thank him for his belief. "A Jedi mind trick, maybe?" "Good man, you're good man," Dr. Mitchel says to Rodrigo. "Let's get you out of here," Rodrigo says fearfully. Rodrigo gets his keys and quickly removes the big lock on the heavy chains around Dr. Mitchel's hands and feet. The rusted chains fall hard to the cell floor. The energy spell created by the rusted metal chains in Gibraltar "the place of the rock" has been broken.

"Thank you so much," Dr. Mitchel says, as his load is lightened. "Much better," he says. They all flow out of the cell and down the isolated cell block. "We don't have much time. Do you have a cell phone on you?," Dr. Mitchel asks. I need to call Birhan in Egypt to tell him that we're on our way." "He needs to pave the way for us and to make everything ready in the chamber." "Yes sir," he says. "You better hurry cause the new shift will be coming on soon," Rodrigo says. They get ready to leave. "Please take me with you, sir." Dr. Mitchel turns and looks at the others. They smile, giving him their approval. "It would be our pleasure, Rodrigo. Let's get out of here. Where are we anyway?" Dr. Mitchel asks. "Gibraltar, Spain," Rodrigo replies "I thought so," Dr. Mitchel says. "I know the best way out of here. Come with me and stay low," Rodrigo says. "We can take my van, it's in the garage." He unplugs the camera system. They all stay very low to avoid detection by the other inmates and guards. They quietly make their way to the parking garage.

The group escape looks funny, but they all make it out. They load up and all fit in his yellow 1968 Volkswagen van. "It had to be a yellow van, no one should notice the little yellow van," Alfonso says sarcastically. "Ok, you got jokes right now," Rodrigo says. Alfonso smiles. As they drive out of the parking lot, Rodrigo sees a fellow guard driving into the lot. He's early today of all days. The relief security guard's eyes get wide when he sees Rodrigo's van full of strange, wide-eyed men, women, and kids. "Oh man, he's early today; that guy is never early. Today of all days he's early, just my luck," Rodrigo says. "Everybody look natural; look natural; nothing to see here, just an escape going down," Rodrigo says. "Wave," he says. They all wave hello and goodbye. He looks confused but continues to drive into the parking lot. He will figure it out in a minute or two. "Where are we going?" Rodrigo asks. "The Gibraltar International Airport. Go! Give me your phone again." Dr. Mitchel instructs. Rodrigo hands it back to him. "Just keep it," Rodrigo says.

He calls Birhan. We got detained in Spain, but we got out. We are heading to the Gibraltar International Airport. "Spain?" Birhan asks. "Yes Spain, it's a long story. I need your hacks to take us off the grid and make travel arrangements for us. Send the E-ticket to this number. We have no passports and we're probably going to be chased. I need you to work your magic brother; there will be nine of us." "I've got you, don't worry," Birhan says. "Give me like 30 minutes. My magic is strong. I have connections everywhere, no problem. Are we still going with you? Yes, of course, you and your family but no one else Birhan." "I got it. I will make sure you get out," Birhan says. "Just get to the airport and I will have my friends there to meet you and take you on their private jet directly to Cairo. The chamber will be ready for you." "Don't forget the stones," Dr. Mitchel says. "I won't," he says. We have about 10 hours to do this or it will all be for nothing. I'll see you in Cairo." "Thank you, my friend," Dr. Mitchel says, then hangs up. Birhan begins to make things happen on his end.

Dr. Mitchel and the others make their way down the street headed for the airport. They all talk among themselves to pass the time and to relieve the anxiety and tension of the moment. Angelia looks at her father. "Dad," she says to Dr. Mitchel. "Yes," he says. "Please tell me how it all began between you and mom." "Right now?" he asks. "Yes, right now. I've waited my whole life, dad." "Ok baby, you're right." He tells her as much as he can as they speed down the Spanish streets. She relishes her father's every word. "I better slow down so we don't get pulled over," Rodrigo says.

They see the airport up ahead. I promise to tell you more as soon as we have time. Angelia nods. "Hey, there it is, slow down," Dr. Mitchel says. "Oh no," Rodrigo says, "They set up a checkpoint with the police and barricades. They must be looking for us," Rodrigo says. "Go around that way," Dr. Mitchel says. " Everybody duck down," Rodrigo says in a panic. They go around the roundabout with their heads lowered between their legs. Sammy somehow doesn't get the message and still has his head up, complete with wide eyes.

"Turn left onto Avenue Del Ejercito." As Rodrigo goes left, Dr. Mitchel calls Birhan. "Hello," Birhan says. "They knew we were coming; we need another way out of here," Dr. Mitchel says. "I figured that might happen, I have a second option", Birhan says." Keep heading East toward the shore. Go to Pier F. My friend Rafael will be there waiting for you. He owns a small dive shop in Puerto, Atunara.

It looks like a warehouse on the outside, It's a perfect cover. Do exactly what he says." "Ok," Dr. Mitchel agrees. At the checkpoint an alert police officer sees a yellow Volkswagen van turning left and calls it in. Alpha 5. Go ahead Alpha 5. I observed the vehicle that fits the description, it turned left at the roundabout and went Eastbound on Ave. Del Ejercito. "Roger that," the sergeant says. He sends three of his nearby officers to go and check it out. The three police officers head out to find them with their sirens blaring. Dr. Mitchel and the others cautiously drive up to the warehouse on Pier F. They see a man standing in front of the warehouse, waving at them in a friendly manner. Rodrigo pulls over.

Dr. Mitchel rolls down his window. "Are you Rafael?" he asks. "I am. You have good friends. Everyone inside quickly; you had better move your asses. Excuse my language, I didn't see the children. "It's ok," Angelia says. "Get rid of that van," Rafael says. Rodrigo's eyes widen. They all rush inside the warehouse, while Rodrigo, Louis, and Alfonso push the van off the dock and into the water. Rodrigo looks sad to see his baby go. It slowly sinks into the water. They run into the warehouse and shut the roll-up door behind them. "Now, what?" Alfonso asks. "Here, put these on," Rafael hands them each a wetsuit. These should fit you; I have all sizes. "I can't do that, I'm claustrophobic," Louis says. "Then you can stay," Angelia says. Louis struggles with the idea of being deep underwater.

They all put their wetsuits on. Rafael gives them all a crash course on how to scuba dive. There will be a submarine waiting for you just outside the harbor." "A what?" Zoe asks incredulously. "A submarine," Alfonso replies. "This is crazy," Zoe says. "Like I said you have friends in very high places," Rafael says. "It will be a short swim out to the sea, you will be fine," he says. Dr. Mitchel quickly takes his heart medication. "Are you going to be alright Dad?" says Angelia. "Yes, baby, I'll be fine. I think I can make it." They each put on their wetsuits, not really knowing what to expect. Police officers notice the van still sinking into the water, not yet fully submerged. Realizing that it is the vehicle that they are looking for, they radio to the checkpoint. "Sir, this is Alpha 7. We see the vehicle here in the water at Pier F. It's right next to the dive shop." "We copy that; set up a perimeter until reinforcements arrive. Hold your position," the sergeant says. "Roger that sir."

They stand there in their black wetsuits ready for action. "Now help each other to get the scuba gear on while I explain how everything works." reinforcements arrive at the scene including the handling sergeant. At the new command post, he gets on his P.A. system. "Fugitives in the dive shop, we know you're in there and we have the building surrounded. You have no way of escape. Come out with your hands raised in the air." Rafael runs to the window and sees the front covered with police officers. "Oh, shit! Ok, you have to go right now." Rafael runs to the back window and sees that there are no police officers in the back yet. "Ok, run my friends, run to the water like baby sea turtles and head for the open sea that way, go!" They slink quietly into the water and start to swim toward the harbor's exit. The group swims, all linked together as the diver propulsion vehicle pulls them along. Rafael grabs a beer from his tiny fridge and opens it. He then puts his earbuds in. He turns on loud rap music. Rafael calmly sits down on the couch and drinks his beer. The special weapons team makes their dynamic tactical entry. They crash through the front door. "Don't move," the sergeant yells. Rafael pretends not to hear them enter, continuing to listen to his music with his eyes relaxed closed. The Police say hello and wave to Rafael. He pretends to finally see them and acts like he is startled. He slowly takes his earbuds from his ears. "What's going on?"

"Where are they?" Who Sir? Rafael asks. "Did you make an actual visual confirmation of the suspects, or did you assume that they were here?" the Sergeant. "Sir, I saw the vehicle fitting the description sinking in the water near here and I assumed that they were here." "You know what happens when you assume?" he says. "Yes Sir." "Damn it, search this building and widen the search to include all other structures in the immediate area. Do you know whose van that is in the water?" the sergeant asks Rafael. With the straightest face he could muster up, he says, "No sir, I've never seen it before in my life." They search the shop. "Sir, there's no one here but him." "That's what I told you." The group has narrowly escaped. They swim in the direction that

Rafael told them to go. Dr. Mitchel sees the submarine and points at it. The captain of the submarine sees them swimming toward him. "There, there they are now and right on schedule. Open the hatch." "Aye Captain." They enter the submarine. The hatch is sealed and closes behind them. The hatch bay door leading into the craft opens. "Welcome aboard the Titan. I hear you need a ride to Egypt. Make sure you give me five stars." He laughs. "I'm like Uber." A very worn-out and bent-over Dr. Mitchel gratefully shakes his hand and nods yes. Soaked and exhausted they all lay down on the deck. We are nine heavy Birhan and heading to Alexandria. "Then I'll see you there in ten hours." He hands the phone to Dr. Mitchel. "He wants to talk to you," the Captain says. Dr. Mitchel takes the phone. "Birhan, we made it safely. You are a miracle worker."

Global Event
between July 16-17,
(Multiple Time Zones)

The President's hand-selected families, friends, and associates from all around the world go down into their underground escape pods and head to the meet location, all at the same time. Every tunnel was carefully built to one day arrive in Russia, deep beneath the small Diomede Island. Diomede, located in the Bering Straits, was the selected location because it is where the greatest portions of the world's landmasses come together. The rest of the world continues to go on like any normal day. The President gets the call while vacationing at Camp David. "Let's go, honey," the President says to his wife Gladys. Pherous says it's time to go." "This is so exciting. What about everybody else Randolph?" "I can't worry about everyone else, our staff, the rest of the family. I'm only worried about us. Our parents and a few relatives are going, that's it. It can only be the ones that I have chosen.

I'm sorry about the rest. Now hurry. Get the kids ready." She does as instructed. They go down into the basement and enter a secret code into the wall panel. The wall moves back and reveals the pod. They strap in and each of them takes their sleep pills. The President, his wife Gladys, and three children, Gregory age 12, Mark age 8, and Sophia age 5, quickly drift off to sleep. "It won't be long now," the President slurs as he falls asleep. Once the correct sequence of numbers is entered into the escape pod's onboard computer, they all jettison down into the thousands of miles of underground connecting tunnels at supersonic speeds. The chosen from all over the planet begin to converge on Diomede. In a matter of hours, all of the thirteen families, charters of the chosen, will all be there together for the first time.

Sabancaya Volcano, Peru
July 16, 2021, 8:39 PM (ET)

Pherous gets a message from the President that comes straight into his mind. Another gift is given to the President from the ball of seed light he was fed in Area 51. "They got away Sir.

"They did what!" Pherous yells out! He cannot be allowed to get to the King's Chamber. I will go to Egypt and take care of them myself. "You and I will have a face-to-face talk very soon I think, " Pherous says. He and his two personal Adimowan guards take sips of their earth-brown hard whiskey from a bucket that looks tiny in their massive hands. They look up as if looking to Adimowa, as though they can still hear the voice of the Highest. They each pour out a little of their drinks onto the ground in honor of the Fallen, an old Adimowan custom that is done on earth too. "I really like this drink," Pherous slurs.

The muted guards nod in agreement and smile. Pherous' halo, on the back of his head, begins to light up brightly. His royal guards begin to light up too. "Take that knife and cut this thing out of my head right now," he orders his guards. They look at him like he is insane. "You heard me," he yells as he takes another swig of his drink. "I hate this thing and what it stands for." One of them takes out his

Adimowan knife and walks around to the back of Pherous' head that still glows brightly. The very nervous guard braces Pherous by grabbing onto his head with one hand as he prepares to remove Pherous' light with his knife. The other Royal guardsmen look on anxiously, having never seen anyone ever remove their light before. Pherous puts his head forward. Suddenly Pherous raises his hand and yells out, "Stop!" The guard stays his shaky hand, having just come dangerously close to starting the removal. "I actually heard this song the other day and it's perfect for this moment," Pherous says, seemingly intoxicated. Pherous begins to sing, "This little light of mine, I'm gonna let it shine." He takes a final sip, finishes his drink, then throws the empty shot glass at the wall. The Royal Guardsmen finish their drinks too. "Let's go," Pherous angrily says after tossing the table across the room. I will free my Father very soon and he will show this world and Adimowa who the real highest is.

Inside the Titan Submarine

Between the Alboran and the Mediterranean
Seas

July 16, 2021, 9:17 PM (CET)

Dr. Mitchel and the others sit in the
submarine's tiny galley area watching a live
news feed on a satellite television hanging in
the corner of the ceiling. "This is a special
report," the newscaster says. They all look up
and start to watch. On the television a seal of a
nation they have never seen before. It's not the
great seal of the United States of America, nor
is it of the United Nations. "We interrupt this
program with a special report; this is a global
emergency. Instructions will follow." All of the
adults naturally just quietly stare at the
television. The kids are still clamoring and
talking. "Quiet," Dr. Mitchel says, "This is it."
Everyone stops and listens. "The following
message is transmitted at the request of the
International Union of Governments. This is
not a test, I repeat this is not a test. Thousands
of people worldwide have gone missing

without a trace. Last night it was reported that thousands from all over the world have been identified as missing. Some people believe this to be the event that the bible refers to as the rapture, or the great catching away.

Church officials say that this is not the rapture because they are still here. This is something else. What's left of our government officials are working to calm the masses, as riots and looting erupt around the world. They made their move to Callisto. They've started the new world and we weren't invited to the party. They must be gathering someplace, or they might be on their way already. "How do you know that dad? "Why are we here, why is God allowing this? I just don't understand?" The newscaster comes on again with an update. "This just in; it seems that many of the missing masses have a common tie. No, it's not religion as some might think, but get this, it's money and status.

People like the Rockefellers, the Rothschilds, Elon Musk, the Hiltons, Bernard Arnault, Jeff

Bezos, Warren Buffet, Mukesh Ambani, Bill Gates types. Of course, the President of the United States, his family, and close cabinet members, just to give you a few examples. Again, this is a worldwide event and you and I have apparently been left behind. Where did everyone go? That is the big question today. Representatives from the International Union of Governments have requested that you please try not to panic and to please remain calm." "There have been worldwide disappearances, but the report is the same everywhere we have checked, the very rich and wealthy and connected have been taken or taken off to somewhere. Perhaps they know something that we don't. Reports are still coming in.

What can we do? God. Do you believe in the Bible? Do any of you believe in it? I never did, it went against everything. I was a man of science. I had to test and prove everything. I think we have been chosen just like Noah, Ezekiel, or Daniel from the Bible. God always has a plan: it's just that we don't always understand it. We just have to go by faith." They all look at Dr. Mitchel in wonder, like he was Moses. They begin to realize that they are on a quest of biblical proportions. They all hug each other. A sound rings out. "Now hear this, we have reached the port of Alexandria; prepare for departure ladies and gentiles. Make ready to surface." "Aye, aye Captain. Birhan will be waiting for you at the Alexandria Aquarium. This way your wetsuits won't stand out as much." The Titan starts to make its ascent to the surface.

Diomede Island, Russia/United States
July 19, 2021, 5:37 AM (AST)

All of the chosen families going to Callisto have arrived at the gathering place deep under Diomede Island. They are getting final instructions before leaving for good. The President addresses the crowd. "We have done it. This moment in time represents the crowning achievement in man's history.

Beyond the explorations of space, we have now broken through the limits that once held us back. In just a few hours we will begin the wondrous journey to Callisto. Please take your sleep pills now; they will take effect in exactly 30 minutes' time. My dear citizens of Callisto, welcome. The large crowd all claps. A team of workers begins to direct the crowd into the hundreds of spaceships all lined up in a row. They prepare each of them for departure. Once the pills have been taken and start to take effect, two of the workers begin to seal each of the passengers in their sleep area.

The rest of the workers go aboard. They enter the codes into the consoles and then buckle themselves into the seat. They take their pills and drift off to sleep as the hundreds of ships begin to take off. They all fly towards the mouth of the subterranean cavern, following their programmed flight pattern. A formation of hundreds of white ships all make their ascension into the sky. The booster rockets all kick in, and they climb higher and higher. They break the sound barrier and enter the stratosphere.

Pherous kneels down on both massive knees. "Father, the time has come! I will avenge you this day and none shall keep me from it. I will make war with the Highest, as you once did. This new Earth will finally be yours as it should be. They will all shudder at your presence again, for I have set you free this day." Pherous rises with renewed power and strength. He powers up and energizes the fire cloud heading to the Great Pyramids.

CHAPTER 11

GO TO HER

Birhan waits at the dock of the Alexandria Aquarium with manager Nicolas Ahmad. They see the group all rise to the surface. They look surprised to see so many people are with Dr. Mitchel. One by one they exit the water. "I have fresh sets of clothing for everyone, I hope they fit you, Nicolas says. "Wow. Welcome to Egypt my friends, I'm so happy that you made it out safely," Birhan says. "Come this way quickly." They are taken to a private area within the aquarium offices to change. and start to take off their gear. " Thank you for all your help, Nicholas." "It was my pleasure," he says. Birhan has a sad look in his eyes, knowing that he

may never see Nicholas again. "Let's go.
Everything is ready." They all jump into
Birhan's black Mercedes van. "We will need to
first go to my house if that's ok with you.
Fatimah and the kids are there waiting for us
and ready to go." "That's fine, but we really
must hurry, my friend. I think that we are
cutting it close," Dr. Mitchel says. "Absolutely,"
Birhan says. He gives a thank-you wave to his
old friend Nicolas.

As Birhan drives, Dr. Mitchel and the rest of
the group of now complete believers take a
moment to rest. They go to Birhan's home
which is located at Abou Bakr El-Sedeek,
overlooking the Eastern cemetery. It's not far
from the Giza Plateau. The children and most
of the adults quickly fall asleep during the
ride. Dr. Mitchel and Rodrigo keep the watch.
"May I ask you a question?" "Yes please do," Dr.
Mitchel says. "Why Egypt? I always felt that
the pyramids were a very special place, but
why are we going there exactly?" Rodrigo asks.
"Let me show you something." Dr. Mitchel

takes out his old journal and opens it up to the pages that will answer his question. He shows him pictures and drawings showing Egypt's longitude and latitude and how all the surrounding rivers in Egypt seem to come together as one into the Nile. "Let me show you something." The pyramids are one of the oldest structures still standing, made about 4500 years ago in 2550 B.C. To this day Egyptian historians say that the great pyramids were built by the help of Shepard-Prophets who came to them." "Visitors?" Zoe asks. "Yes," Dr. Mitchel says. "While I was with Makia she taught me that the pyramids were built by Egyptian Pharaoh Cheops of the fourth dynasty with help of the Prophet Enoch." "We're here," Birhan says. Everyone starts to wake. "To be continued," Dr. Mitchel says. Rodrigo is truly fascinated.

They all get out of the van and look around as they dart straight into Birhan's house. "Make yourselves at home. I just need to make some final calls to pave the way for us," Birhan says. Fatimah and the girls greet and meet everyone. "Thank you so much, Horace," Fatimah says. "No, I should thank you and Birhan for allowing us into your home and the pyramid." The kids all start to gravitate to each other and start playing. Birhan comes back into the room with a wooden box in his hands. "Is that what I think it is?" Dr. Mitchel asks. "Yes, it's the stones. The very same ones that were in the breastplate." "Perfect." Birhan and Fatimah leave the room for a moment, while everyone else hangs out. Rodrigo looks over to Dr. Mitchel. "You want to know more huh?" Dr. Mitchel asks. "The truth is addictive. You left me hanging Sir," Rodrigo says. Dr. Mitchel smiles. "Enoch was Adimowan, one of the first visitations to take place. The very same Enoch from the Bible. That's why the area just before you enter the King's Chamber is still known as Enoch's circle. It is 365 degrees, a perfect circle, and the same age as he was when he was taken back to Adimowa. The Bible is very clear about this. "Birhan, do you have a Bible?"

Dr. Mitchel asks. "Yes, of course," Birhan says. Fatimah gets it from his study. "Here you go." "Thank you," he says. "In the book of I believe Hebrews, yes the new testament book Hebrews chapter 11 verse15." Birhan quickly finds the scripture in his Bible. Please read that." "It says, "By faith Enoch was translated (changed, transformed, transported) that he should not see death; and was not found, because God had translated him: For before his translation he had this testimony, that he pleased God."

"Thank you Birhan. You see, he never saw death as a man. This proves that it is possible. Translated by God simply means that he was changed somehow and taken back to Adimowa. It's important to note that he pleased God." Wow, I never knew that translations were in the Bible," Rodrigo says. "Yes. You see, you must imagine in your mind's eye Enoch standing at the base of the great pyramid with his Adimowan measuring devices in hand, as thousands of Egyptians move the great stones with lost Adimowan antigravity knowledge. Pharaoh prepared himself in the King's Chamber, trying to discover the correct incantations to translate. He possessed ancient knowledge that he built into the Great Pyramid by order of the Highest." "May I interject?" Birhan asks. "Absolutely," Dr. Mitchel says. "It is what the ancient Egyptians called a "Peq"; that's a place where the dead enter the afterlife.

The great pyramid was built at the exact center of the planet for the purpose of, what we know now, teleportation to other realms." "Exactly," Dr. Mitchel says. "It is the only place on Earth where can be translated while still alive to Adimowa, with your human essence still intact." Wow. They all look amazed. "The Egyptian pharaohs believed this with all their hearts. We must get to the King's Chamber and pass-through Enoch's circle, climb what was once known as "Jacob's Ladder." Thank you, Dr. Mitchel, that was very detailed and eye-opening, Rodrigo says. You're very welcome, says Dr. Mitchel. Louis, who has been listening to Dr. Mitchel and Rodrigo the whole time while resting his eyes, pulls Dr. Mitchel away from everyone else and says, "I only have one more very important question left for you that Rodrigo forgot to ask you, Sir." Dr. Mitchel looks at Louis, knowing that it will be some comedy involved. "Go ahead," Dr. Mitchel cautiously says. "How was it?" Louis asks. "How was what, Dr. Mitchel replies?" "You know," Louis says. Oh. "Gentlemen never tell Louis; I'm old school like that." "Come on, this is different Doc," Louis says almost begging. "I will say this. Dr. Mitchel pauses for effect. "it was out of this world Louis." They both laugh. "That was a good one," says Louis. "We seriously need to have a real talk about that when this is all over." Dr. Mitchel smiles.

Birhan smiles at Dr. Mitchel. "Thank you so much, my friend. This means so much to me and my family." Just then, the front doorbell rings. They all stop and look at the door, wondering who it could be. "Fatimah, are we expecting anyone?" Birhan asks. She looks like she knows something but is afraid to tell. The door rings again. "Open it," he says. Fatimah walks over to the door and looks through the peephole. She looks back at the group then opens the door. "Hello!" they all say. Standing in the doorway are half of Fatimah's friends and some more of her family. Dr. Mitchel looks at Birhan. "I told you no one else," he says. "Fatimah!" Birhan screams. Fatimah looks over at Birhan with the most innocent look and gives him the eyes. "We were just leaving," Birhan says to the group with an uncomfortable smile. The mob almost forces their way into the house and greets Fatimah and the kids. Suddenly it's a party. Fatimah has let her emotions get the best of her and invited everyone she knows to go. May I see you outside Birhan?" Dr. Mitchel asks in a firm tone.

They both step out onto the front porch and close the front door. "I will get rid of them," Birhan says. "I wish we could take the whole world, but we can't brother. There just isn't enough room or stones for that matter," Dr. Mitchel says. "I know," Birhan says, "but they probably won't take no for an answer at this point. If we turn them away they may even alert the authorities. See that large man over there?" Dr. Mitchel glances over his shoulder. "Yes," he says. "That man is a cop, actually a lieutenant. Right now, he's on our side," Birhan says. "Ok, let me think," Dr. Mitchel says. "Put the stones in the van and lock them in. We'll let them think they are going for the moment. Just follow my lead and we'll fix this.

"Birhan and Dr. Mitchel go back inside and shut the door behind them. "May I have your attention; everyone please gather around," Birhan says. This whole time Fatimah has been doing her best to feed everyone. They all stop what they're doing and turn their attention to Birhan and Dr. Mitchel. Dr. Mitchel steps forward. "This is the world-renowned Astrophysicist, Dr. Horace Michael Mitchel," Birhan says. "He is the one who discovered what the Great Pyramid was designed for."

They all start to clap their hands and some even start to cry out for joy. Dr. Mitchel smiles and is truly humbled by the display. Birhan points his open palm at Dr. Mitchel. "Please say something Doctor," Birhan says. He is surprised that Birhan asked him to speak but remembers that they must put on a good show. The Police lieutenant looks at him with eyes full of hope. "Thank you all very much. I know that you are all here because you have learned what is going to take place very shortly. They all nod yes. Honestly, I don't know if any of this will work." They all start to look at each other. "The theory that I've come up with is still just that, a theory." You can feel the air being sucked out of the room.

Faces of happiness quickly turn to faces of surprise and even anger. Dr. Mitchel continues on with his charade. "I'm so glad that you have all come here this evening and have decided to invest in our little venture. They all start to look around and wonder what the heck he is talking about. They start to whisper and talk among themselves.

They all look at Fatimah. She raises both her arms and gestures that she doesn't know. "What is this?" one of Fatimah's friends asks. Dr. Mitchel can feel the atmosphere in the room change but he doesn't stop. "I will get right to the point because I know that your time is valuable," he says. Zoe, Alfonso, Louis, Angelia, Rodrigo, and even the kids are confused by the show they're putting on. They start to figure out that it's a plan to get the extra people to leave. They start to play along. "Dr. Mitchel, would you like me to start up the two-hour PowerPoint presentation?" Zoe asks. Dr. Mitchel stutters yes. "But before you do, I want to now give our potential investors here, the ground floor opportunity to invest right now. We will only be accepting cash investments in the amount of $40,000 American and higher. Several people tell Fatimah what they really think and leave. Several more angrily storm out, wondering what they got themselves into. One by one they leave. The master plan appears to be working. All but one person is gone. Nasim, one of Fatimah's friends and co-workers, remains seated on the sofa. He stands up and starts to clap his hands. "Very good show Dr. Mitchel.

I know that this was no investment party and I know that what Fatimah told me is true. I want to go; I need to go with you." "We cannot take anyone else, Sir. We simply do not have the space," Dr. Mitchel says. "I will tell the police if you don't take me," Nasim says. They all look at him. Louis, Rodrigo, and Alfonso move in and slowly box him in. "How much is it worth to you?" Nasim asks. "I really wish you hadn't said that," Dr. Mitchel says. They grab Nasim and take his cell phone away. He looks surprised and a little scared. He never thought that it might come to this. "What are you doing?" Nasim asks. "We need to go now; we're already behind schedule. I wish we could take you Nasim, but we simply can't. I'm sorry to have to do this to you," Dr. Mitchel says. "Tie him up and put him in the closet." Fatimah looks very surprised to hear Dr. Mitchel say that, then helps to tie him up and put him in the closet. "What are you doing?" he asks.

They give him food and water and a very dull butter knife from the kitchen. From inside the unlocked closet, he screams for help at the top of his lungs. Birhan says, "Don't you just hate when people show up unannounced?" They all laugh at his joke. "He'll be alright and will eventually get free. We need to tape his mouth, someone might hear him." Birhan gets the duct tape from his tool drawer on the back porch. He opens the closet door. "I'm so sorry my friend." Nasim can only yell in a muffled sound. "Let's go," Dr. Mitchel says, "we have no time to waste." They all load up in the van and take off for the Giza Plateau.

In the distance, the Great Pyramid can be seen. It is a stunning and awesome sight to see. It's very beautiful even now but when it was first built it was overlaid with white granite that reflected the sun from all sides. It could be seen for miles. "Wow, it's just as I imagined. I have always wanted to come here. It's one of the places on my bucket list," Zoe says. "Well, you are about to get the tour of a lifetime," Dr. Mitchel says. They arrive at the base of the Great Pyramid, and they all get out and stretch their legs. "Wow Pawpaw. Look, mom and dad, it's so big. I can't believe it," Sammy says. They all peer out the window of the van as they each exit. "Come, it will take us some time to reach the King's Chamber. We must hurry." Birhan speaks to the guards posted at the perimeter checkpoint and gives them a large roll of cash. They allow them to move forward toward the base. At the base, he does this again and it works. They now have complete access to the Pyramid. The group makes their way into the Al-Ma'-Mun robber's entrance and begins their ascent to the King's Chamber.

They all crouch down and enter the shaft. Right at that moment, Pherous arrives in his fiery cloud. Dr. Mitchel looks behind them and looks up into the sky. He sees the redness of the fiery cloud in the distance and notices that it is moving closer. "Go faster," he says to the group. "Strange weather we're having, no?" Birhan asks, knowing that something is not right. "Yes, very strange; we need to move," Dr. Mitchel says. As the group goes into the entrance, Pherous can see them. Pherous is still a good distance away but can see them just the same. Because of the energy that resides in Egypt, Pherous is unable to show up instantaneously where they are. The guards see his fiery cloud coming towards them and fall to the ground, gripped by fear. They start to pray to the East thinking that Allah is angry for what they did. Dr. Mitchel happens to turn just enough to make out Pherous exiting his cloud. He remembers a bible scripture that describes this. Revelations 1:7 "Behold he cometh in the clouds, and every eye shall see him."

Pherous makes his way towards the entry shaft. It takes Pherous several minutes to even get to the entrance of the main passage. The group has a head start but Pherous is gaining on them fast. Dr. Mitchel and the group continue to crouch down in the passage going towards the King's Chamber. Dr. Mitchel quickly thinks up a plan to buy them some time. They tell the group to continue climbing forward as they turn back to set a trap. "Here, Birhan, help me throw some of that old lamp oil onto the smooth limestone floor, right near the entrance." "You think that will work?" Birhan asks. "It's all we' e got." Birhan helps Dr. Mitchel to throw the old lamp oil all over the smooth limestone floor near the entrance of the first ascending passage, at the granite plug. "That should do it. He won't be expecting that," Birhan says. "Simple, yet effective, I hope," Dr. Mitchel says.

Pherous reaches the entrance to the descending passage. He crouches down, but it is very tight for him. As Dr. Mitchel and Birhan make their way up the ascending passage to catch up with the rest of the group, Dr. Mitchel grabs his right thigh in agony. Birhan looks back at him. "You, ok?" he asks. "It's my leg, I think I pulled something. Don't ever get old Birhan," Dr. Mitchel says. "Too late my friend," Birhan says. Birhan takes Dr. Mitchel under his arm and helps him to keep moving forward. The group struggles as they continue to crouch down and move up the ascending passage toward the Grand Gallery. The children are very tired. Fatimah acts as the guide in Birhan's absence. "It's just a little further until we reach the Grand Gallery; it will be easier to walk when we get there," she says. "Help each other and keep moving forward," Fatimah says. "It smells funny in here," Mona says. Angelia and Louis for once work together to keep the kids together, moving forward, as Alfonso helps Zoe.

Rodrigo leads the way with Fatimah and the girls. It's hot and stuffy. In the distance, making their way to the group is Dr. Mitchel and Birhan. Dr. Mitchel is in a lot of pain, but Birhan helps him walk. "There's Pawpaw and Birhan," Josh says. They are exhausted but they finally catch up to them.

Pherous makes his way down the passage moving toward the ascending passage in pursuit of the others. As they begin to head up the ascending passage, Pherous totally bypasses the oil trap and heads straight down the descending passage that leads to the Subterranean Chamber. He grabs onto the sides of the passage and slides in freefall all the way down. He comes to a sudden stop. He is in complete darkness there. The group finally reaches the Grand Gallery, which is large, wide, and mysterious. It is staggering in its magnificence. They all look in silent wonder as they continue to press their way to the King's Chamber. "I'm tired," says Gabriel. "Me too," says Mona. "Can we please stop?" asks Sammy.

"Ok everyone, rest for a few minutes," says Dr. Mitchel. They all sit on the floor, midway in the Grand Gallery. "Hydrate, that is very important," Fatimah says. They pass out bottles of water from Rodrigo's backpack. "Birhan, did you remember to bring the special water we need?" Dr. Mitchel asks. "Yes, of course," Birhan says. Angelia and Louis help each other and make sure the kids all drink. Zoe and Alfonso rest as they take in the beauty all around them. "I always wanted to see the Great Pyramid and here I am," he says. If you would have told me that in two days you will have been to Spain and then inside one of the great pyramids of Egypt, I would have said that you are crazy," Zoe exclaimed. "Tomorrow we'll be in Heaven."

They both sit and stare up into the Grand Gallery. They hear a loud roar that sounds like a hungry Dinosaur, coming from below them. Pherous blasts away the iron fence that blocks the chamber entrance. He illuminates his body and gives light to the pitch-black space.

"Father!" he yells. He falls to his knees and begins to pound the unfinished ground with his massive fists. He lifts his hands over the pit, palms facing downward, and says strange Adimowan words. "Father," he yells, morphing his voice to sound like an animal. He energizes a ball of red light and drops it down into the pit. It seems to fall forever. The pit is only three meters deep, but it contains eternal properties that make it endless. They all look at each other. "Let's go now, everybody as fast as you can to the King's Chamber!" Dr. Mitchel yells. "Pherous is coming!" Dr. Mitchel figures out that not only is Pherous there to stop them, but he is there to break someone out. He opens his journal and reads from it. "Oh my God, this isn't just a place to translate, it's also a prison," Dr. Mitchell says. "The bottomless pit is here in the Subterranean Chamber." "What's that?" Angelia asks. "It's a place where Adimowa would imprison someone they never intended to be free again. That is the other purpose of the Great Pyramid. I just figured out who Pherous is.

Have you ever read about the fallen Arch Angel named Lucifer?" Dr. Mitchel asks. They all nod yes. "The Devil?" Rodrigo asks. "Yes, exactly. Lucipherous is here, in that chamber. It says here that he was banished to Earth by the Highest and sealed in the great pit, but it didn't say exactly where it was. The bottomless pit is here in the Subterranean Chamber. I forgot all about the Subterranean Chamber." "My Son," Lucipherous says, embracing Pherous. Even the Devil loves his son. "You have done it. How?" "Nothing could keep me from it, not even him," Pherous says. The chosen one is here trying to get to Adimowa." "Where is he?" Lucipherous asks. "The King's chamber," Pherous says. Together they transform into raging beasts. They tear, claw, and climb their way up the passage, destroying the limestone with every inch, from the lower Subterranean Chamber to a good shaft that leads vertically straight up to the ascending passage at the entrance of the Grand Gallery. Dr. Mitchel and the others arrive at the top at the great step. They help Dr. Mitchel to get over the step. He struggles but continues. He can barely take a step and must be carried the rest of the way. Louis and Alfonso carry him in a seated two-handed seat carry. They all finally arrive at the entrance to the King's Chamber. They hear loud roars from below. They lay Dr. Mitchel on the ground. He grimaces in pain.

He pulls out his old leather journal. Again, another loud roar and the ground beneath them begin to quake and resonate with sound. "Everyone remove your shoes and get into the sarcophagus," Dr. Mitchel commands. They look at him like he's crazy for a second, then do what he says. They all take off their shoes and throw them down. "Come here Sammy," Angelia says. They pick up Sammy and Mona and place them over and into it. Josh climbs in; then he helps Gabriel and Dellina to get in. Then they help Dr. Mitchel to get in. He is clearly in a lot of pain. "I need my medication," he yells. Louis helps Angelia, Alfonso helps Zoe. "I can barely breathe and my claustrophobia... I can't do this," Louis says. "Yes, you can," Angelia says. She helps him to get in. Birhan and Fatimah both get in. Rodrigo hops in last. Somehow, they all fit in the ancient red granite box, standing up. "Birhan, hand me the water." Birhan takes the water from his backpack and hands it to him. Dr. Mitchel fills the floor of the sarcophagus with purified water. The water mixes with the natural salts in the box. "Close your eyes, and don't be afraid. Faith is the key, just believe." More sounds from Pherous and Lucipherous can be heard as the group all shut their eyes tight. Dr. Mitchel opens his journal and turns the pages frantically. "Listen to me, the children under 12 don't have to worry because they are not held accountable yet. Josh and Mona, you must believe. "Believe what?" Josh asks. Believe that the Highest is real, you must believe for yourselves. "I believe," Josh says. So does Mona. Dr. Mitchel begins to read from the old journal various words, following Makia's instructions to the letter. Birhan has a tremendous smile on his face as he holds Fatimah and his girls.

Pherous and his father's roars sound very close. They can hear the destruction and power moving through the solid pyramid. Dr. Mitchel and the others look at the small door to the King's Chamber with wide eyes. He tells Louis to remove a small brown pouch from his jacket pocket. He takes the pouch. In the pouch are the twelve precious stones needed to translate to Adimowa. Jasper, Sapphire, Chalcedon, Topaz, Beryl, Sardonyx, Onyx, Amethyst, Agate, Emerald, Amber, and Flint. Dr. Mitchel has a hard time getting his words out clearly. "It says to take a stone in your right hand. Wait a minute, oh my God, what have I done? They all look at Dr. Mitchel. There are thirteen of us and we only have twelve stones. There can only be twelve! Pherous' and Lucipherous both roar like wild beasts. They have gained some power coming out of the ancient pit. They position themselves in the ascending shaft that leads straight up to the Grand Gallery. Pherous lifts his hands and lets out another ear-defining roar. Large ugly claws come out from his fingertips. They both leap into the air and start to claw their way up the shaft with their renewed power. They are moving fast to the top of the shaft, roaring like lions on the way to the Grand Gallery. Dr. Mitchel hands them each a stone from the breastplate of the priest but he doesn't have one left for himself.

"Oh shit, I messed up. It says it takes "twelve stones from the breastplate, for twelve souls to translate." "Ok, you guys go on without me!" "No Dad, you must come with us!" Angelia says. "We can't do this without you, I love you so much! I can't lose you now!" "Pawpaw no, you can't do this, we need you," Sammy says. They plead with Dr. Mitchel to go with them, as Pherous and his father are getting closer by the minute. None of them have noticed that

Rodrigo has climbed out of the sarcophagus and is standing by the chamber door watching them. He gets their attention by clearing his throat and waving. They all stop and look over to the entryway. "Rodrigo, what are you doing? Birhan asks. "I'm a believer. Hey! I'm the only one that's trained for this one. I'm a true believer, a Judean...sacrifice is in my blood." "There must be another way," Dr. Mitchel says. Pherous and Lucipherous are almost there. "No time Sir. Since we spoke at the jail, I knew then that this was the reason why I was born. This is the right thing to do. I believe that all this is true, and I know that you will get them there. The sincerity in his eyes said it all. "I'll go and buy you some time. I got this. I'll see you on the other side family."

They all stand silent, tearful, and grateful to Rodrigo. "Go and tell my Dad I said I'll see him very soon. Go!" Rodrigo throws the jasper stone back to Dr. Mitchel. He catches it and then stretches his hand out to Rodrigo and looks at him. He points to the sky and Rodrigo nods in affirmation. He crouches down and heads down the dark passageway from the King's Chamber and back down the Grand Gallery. He can hear the roars as he comes up the shaft from the Subterranean Chamber. Rodrigo stands ready to face the Devil and his son. Dr. Mitchel and the chosen twelve now stand ready to go. He gives them more of Makia's incantations from the journal. "Take the stone in your right hand and place your closed fist on the center of your forehead and hold it there," Dr. Mitchel says in a very weak voice. They all do it. He reads the final ancient words from the book and initially, nothing happens at all. They wait and stand in the silence and darkness, stones to the forehead, feet wet from the purified water and salt. More roars in the distance, but getting closer. They anxiously wait for something to happen. "We're all going to die," Alfonso says to Zoe. "No," Zoe yells, "it's going to work, just believe!" "Did you write it down right?" Louis asks. Dr. Mitchel is silent. Smoke and limestone blast from the passageway into the air.

Pherous and his father Lucipherous have made it all the way into the Grand Gallery. They rage blast out of the shaft and land on their feet together. Their roars echo and resonate through the Grand Gallery like Gregorian Chant. Rodrigo prays and then takes a fighting stance. He waits for his destiny to arrive. He looks up and sees both great black beasts. "Let's go! he says. Rodrigo starts to dance around like a boxer getting ready to rumble, "Lord, give me strength," he says. Pherous walks over to and changes back into his rather large but more human form. "I know you, guardian of the King's Chamber, protector of Adimowa's secret, Rodrigal." Rodrigo looks surprised that Pherous knows his name. Lucipherous stands back to watch his son let loose on the weak human. "Let's get this over with, shall we?" Pherous says. Pherous takes his fighting stance and approaches Rodrigo. It's like David and

Goliath all over again. Rodrigo and Pherous engage in a fierce battle. "I will see you destroyed by the Highest." Rodrigo rushes in then runs back out. He tries to punch Pherous in his stomach but misses. Pherous gets a flash of light in his eyes. "This place throws me off," Pherous says. Rodrigo looks back at the entranceway to the King's Chamber, hoping that he has stalled Pherous just long enough for the rest to get out. Pherous moves quickly and cuts off Rodrigo, pouncing down on him like a jungle cat. He grabs him by the hair and lifts him high in the air with both hands. Pherous lets out a loud growl, then rips and tears Rodrigo's body in half. He throws both pieces against the hard limestone wall as if he were throwing out the trash. You can barely make out that the pile of carnage was ever a man. Lucipherous transforms back into his human-like form. He starts to clap his hands in utter delight. "No mercy, I love it," he says. Pherous and Lucipherous turn towards the King's Chamber with no one to stop them. The

group   is   still   standing   in   the
sarcophagus holding their stones against their
foreheads. They hear a small faint sound and
begin to feel an energy surge through the air
of the chamber. A rumble can now be felt
coming from deep within the pyramid. A loud
sound begins to resonate throughout the
pyramid, like the sound of a great trumpet.
Each one of their stones begins to light up
while on their foreheads, illuminating their
faces where they hold them. The ground
begins to shake and move under them.
Pherous and Lucipherous both smash into the
King's Chamber. They are barely able to stand.
The energy and power of the Adimowan
incantation fill the room and intensify with
each passing moment. They fight to make
their way to the sarcophagus to try and stop
them. They cannot stand in the presence of so
much conjured-up Adimowan power. They
must bow down to the ground as they are
pressed to the ground. Colored light shoots
from the mouths and eyes of each person. "It's

happening!" Angelia shouts. Their hair is electrified and stands on its edge. They begin to levitate above the sarcophagus. In the twinkle of an eye, they all disappear from their human form and turn into hovering-colored balls of light, each one in the color of the stone that they were given. Now colored balls of light, streams into a line of brilliant light, as one.

In a beam of light, they shoot up through the granite slabs, straight through the top of the King's Chamber and out of the top of the Pyramid. Faster than the speed of light, the beam of light steaks through time and space, to eternity and the edge of everything. The group still has a sense of their own consciousness and has an awareness of their person. They look around at each other's essence of being in the beam of colored light, not afraid and at peace as they travel through the fabric of time and space. By the awesome power present in the King's Chamber, Pherous and his father get vacuum- sucked out of the

King's Chamber, down the shafts, and spat out of the Great Pyramid. They land hard on the desert floor like a meteor, leaving a large crater. Like a jackal with its tail tucked between its legs, Pherous and Lucipherous scurry away into the desert night. Pherous stops and looks to the sky. "Go to her, for surely, she waits for you! Tell the Highest I am here, we are here waiting for him! Bring it."

Lucipherous looks at his son with devilish pride.

CHAPTER 12

UNLEASH HELL

WASHINGTON, D.C, THE WHITE HOUSE JULY 17, 2021 5:28 PM (ET)

A small group of White House staffers "shelter in place," in one of the last safe locations in the country. Desperate people smash their way through the gates and break into the lobby area of the White House, ransacking and desecrating it as they move through it. There is a Benghazi incident atmosphere that was never thought to be possible in America. They make their way up to where the White House staff, silently waiting in the West wing. "Maybe we should just let them in," Dianne Moore says. "No, are you crazy?" Mark says. "They just left us here, after all those years of service to them," Dianne says. "We were just the help, nothing more to them," Mark says. "They are going to get in, we need to move now." Fearing the mob that is about to come through the door, they move into the Oval Office.

They enter the sacred office area of the President who is now long gone. The hungry mob breaks through the outer office area and then comes straight to the Oval Office. They are unable to lock the door behind them. "There they are!" one of the protestors says. The staff members are overrun and brace for whatever is in store for them. They now stand face to face in a kind of standoff. The insurrectionists are surprised and almost disappointed to only see low-level staffers and interns. "They're all gone," a protestor says sadly. Their anger subsides and deflates like a balloon. They unclench their fists and drop the objects in their hands to the decorative marble floor.

"Where is the government, where is the assistance, where is order?" The staffers and interns are now a part of the desperate and angry, seeking answers. "Oh my God, look," Dianne says. They all stop and stare at the flashing red light coming from the desk. The red button has been activated in an opened briefcase under the desk. It is a nuclear countdown to the end of the world. They all stop in their tracks. "Those bastards, we're screwed, they're going to kill us all," Mark says. They all walk away realizing the severity of the situation and the stupidity of what they were just doing to each other. "It's over, everybody go home to whatever family you have. It's over," Dianne says. They all turn and go home to their families and loved ones with the horrible, dismal truth that they are all going to die very soon.

Once people all over the world start to realize that the reason why the rich and politically connected have disappeared is that the world is coming to an end, it's not long before hysteria breaks out. Whether this is true or not, the prevailing thought spreads like wildfire through the hearts and minds of old and young. The major cities quickly fall into dissent and chaos. Riots, fires, and looting can be experienced everywhere. It spreads like a global pandemic but there will be no flattening of this curb. Militia groups have taken this opportunity to release their plans for anarchy. Terrorist organizations have used it to accelerate their nefarious plans.

Various like-minded individuals start to form groups, which feels a lot like going back to a tribal system. The stock market is completely closed and assumed to have crashed. Computer systems have failed, disabling our so-called technology overnight. No planes fly overhead, no tv programming, nothing. Port operations have come to a grinding halt and have backed up leaving ghost ships all over the harbors. Trucks and trains with needed food transports have stopped. Grocery store shelves go empty and will not be filled again. There is no leadership in the world, no White House press conferences to calm the masses, no address from the United Nations.

There is no martial law implemented by FEMA; no curfew. There is only the law of the jungle, kill or be killed, survival of the fittest. Every man for himself. The plan was the complete breakdown of society, the destruction of all that was left behind, seemingly the start of the biblical great tribulation. Literally, all hell breaks loose. The plan was executed flawlessly.

United States Pacific Fleet
Somewhere in the South Pacific
July 18, 2021, 2:19 PM (PT)

Naval Captain of the U.S.S Theodore Roosevelt Aircraft Carrier, Captain Larry Aldridge reads from the teletype in his office dispatched from the Office of the President. He cannot believe what he's seeing. He has just received and confirmed the orders to launch a full nuclear counter strike. Missiles inbound. Nuclear war is imminent. DEFCON 1. He quickly hits a button that activates an alarm sound from the bridge. "General quarters, all hands to battle stations!" go out over the PA. Captain Aldridge starts the launch sequence by inserting his special key. Sailors frantically run from every part of the ship to get to their assigned stations. This is not a drill. The rec rooms, gyms, galleys, and bunk rooms go empty.

Chief Officer French stands by the captain. "Mr. French, we have enemy missiles inbound, attack is imminent, we're at DEFCON 1. Launch a full counter-strike," the captain says. "Captain," the chief officer says, hesitating. "You heard me damn it!" After understanding the order, Chief Officer French inserts his launch key. They both look at the bridge phone, inwardly hoping to get a last-second override call. "Start the launch sequence now! Verify and confirm the order Mr. French!" "Verified!" he yells. "The order is verified, the order is confirmed, the order is true," Captain

Aldridge says. "Starting, launch sequence now! May God have mercy on our souls."

They both turn their keys in unison. The captain then hits the red button. The mighty aircraft carrier launches all its arsenal of nuclear weapons at the predesignated targets around the world. All United States armed forces launch at the same time, from underground bunkers to bases in South Korea. It has happened, full thermonuclear war has commenced. The nightmare scenario has started around the globe. Nuclear holocaust, the destruction of mankind, civilization as we know it has shifted. The nations with nuclear capabilities do the same and have also been annihilated. Major cities around the world are wiped off the face of the planet in a flash, blasted back to the stone age. Most of the missiles that were headed to U.S. cities were intercepted by counter missile defense systems outposts. The oceans are flashed vaporized and are now for the most part a barren Sahara-like desert. One-half of the world's population is killed and a third will die from radiation exposure in the coming weeks. Where is God, the world wonders? Some see it as judgment day, a day that we collectively have deserved for some time now. Some wonder how we ever lasted this long. Some see it as a prophecy fulfilled. The Day of Visitation, if you will, has come to the Earth.

Nasim finally cuts through the ropes and duck tape that once held him bound in Birhan and Fatimah's front room coat closet. The closet door swings open. "Finally," he says, taking in a deep breath of fresh air. "Fatimah,!" he yells. Even though he is very hungry and extremely dirty, he actually stops and takes the time to urinate on their front room sofa and dining room table. Satisfied with his small dose of payback, he opens the front door and steps out onto the porch. Looking up into the evening sky Nasim sees the first Intercontinental ballistic missile to touchdown in Egypt. "No!," he screams, as he is vaporized from existence.

CHAPTER 13

TWELVE SOULS

DEEP SPACE AND TIME, BOUND FOR THE RIVER OF LIFE, 1ST ADIMOWA

The twelve souls, still radiant balls of colored light, come up through the darkness and beauty of space toward a snaking rectangle above them that gets smaller and smaller as they get closer and closer. It's the to the River of Life, the place where souls are reborn into Adimowa. As their lights come up through the bottom of the water, they each become their perfected essences. They each flow up and into the River of Life, like all other souls that come in daily. The twelve souls have finally made it, they have safely translated and been transported all the way from The King's Chamber in the Great Pyramid to the River of Life in 1st Adimowa.

Even though they are still balls of bright light, they can sense that Adimowa is beautiful in every way imaginable. Crystal-clear waters, green and lush untouched meadows, even trees like the ones on Earth, the most beautiful landscapes ever seen by Earth-bound eyes. Everything is gold-mixed and incorporated in Adimowa's construction of almost everything. Polished gold is everywhere you look. There is no Sun but the light is still everywhere. They have arrived and are floating in the River of Life that flows into the Great Temple of the Lower Lords, in first Adimowa.

The twelve are collected by one of the guardians of souls and gathered but not weighed as usually done. The gatherer has immediately recognized their rareness; he perceives and knows that their number, color, and mineral sequence combined are prophetic and must be handled specially. "What is this?" he says curiously in the Adimowan language. He separates or makes holy the twelve special souls from the other ones continuously arriving. He puts the balls of light in a golden pot and takes them to Makal, the Arc and overseer of Souls. The guardian kneels at the feet of Makal who is dressed in the finest blue glowing garment and sits the pot down on the solid gold floor in front of him. "What is this?" Makal asks. "Twelve stones mingled with twelve souls; it is prophetic Sir," the guardian says. Makal stands up, walks over to the pot, and looks down into it. "We have not had a translation in thousands of years," he says. "Let's see what we have. Pour them out into the pool." The Guardian of Souls carries the golden pot to the regeneration pool on the right of Makal's throne and pours out the twelve souls, still balls of illuminating light. The waters of the pool instantly transform the twelve from balls of light into the perfect Adimowan essence and representations of themselves. The guardian looks amazed to see the new arrivals as he returns the golden pot to the River of Life. Dr. Mitchel stands tall, is thirty-three years old again, and in perfect physical health. He has no more need for his heart medication or any medication for that matter. "I feel good," he says to the others. They feel like fish that have been returned to the water. It is all so perfectly natural and normal. All the children are now twenty years old and beautiful. With their increased age comes increased intellect.

They are fascinated by it all. Louis and Alfonso are also tall and strong, are Thirty-three years old, and have no more pain whatsoever. Their bruises and marks from the beating they received in Florida not long ago, are completely healed. Angelia is tall and Adimowan beautiful. She and Louis no longer hate each other. They feel nothing but love and gratitude. The rest of the group is also thirty-three. This seems to be the perfect age within Adimowa. Everyone is at least twelve feet tall. "We are playing by Adimowan rules now," says Dr. Mitchel. He waves his arm, motioning for them to bow. They all bow down at the feet of Makal, in reverence. "Rise, servants of the Highest. "Do not bow to me. Why have you come the way of Enoch, the way of the ancient ones?" "Makal, Arc of the Highest, we have come to you to seek his everlasting wisdom regarding his creations on Earth," Dr. Mitchel says. Pherous and his father Lucipherous have unleashed their fury on Earth. It has all but been destroyed by nuclear fire.

"I was the chosen one of the visitation in Earth year 1979. I have also come to beg your forgiveness as well. I brought great shame upon you and our family." "Our family,?" Makal asks. "You!" Makal says in a strong voice. "Horus." Dr. Mitchel lowers his head in submission to whatever the consequences might be. Makal looks down at him with an angry and fierce look on his face. "I knew that you would come one day," Makal says. The Highest has spoken regarding you and my daughter a long time ago. "I have come here to find Makia," Dr. Mitchel says. "This is her daughter and your granddaughter, Angelia." "I know. I see Adimowa in her eyes. Come my Granddaughter," he says to Angelia. With tears of absolute joy rolling down her cheeks, she and the children go to him. He touches all their heads and blesses each one of them. Makal looks at Dr. Mitchel and says, "Do you think that my daughter waits for you, through eternity?" "No, but I waited for her all my life, even into eternity where I stand now. I have never loved another. Where is she?" Dr. Mitchel asks. Makal lowers his head and turns away. "She was banished to outer Darkness for 10,000 Earth years, separated from the Highest. The price for your transgression was death." "No," Dr. Mitchel says. "My love, what have I done to you?" He falls to his knees and begins to cry. Makal watches him closely. You know that there is no sorrow or crying here. Stand up," says Makal. Dr. Mitchel stands up. "Can I go to her?" Dr. Mitchel asks. "Yes, you can, but I would not recommend it," says Makal. "No dad, you can't do that!" Angelia yells. "There's no coming back from that Horace," Birhan says.

"I don't care, if she is there, I want to be there with her. May I please speak to the council, something must be done," he sobs. "I believe that you would go to outer darkness for her." "Yes, I would, I love her," he says. "I can see that," says Makal. He looks over Dr. Mitchel's shoulder and smiles. 10,000 Earth years is only 10 days here, Makal says. Makal starts to laugh out loud. They all turn around to see what he is looking at. Makia has been standing behind them listening to his every word this whole time. "That was wrong," says Alfonso. Makia looks at her father, seeking his approval. He nods and smiles. Dr. Mitchel turns toward her and then is frozen in his tracks, in total shock. She is still thirty-three, possessing ever the same beauty. With perfect love showering her every emotion, her eyes fixed on Dr. Mitchel, her love. She lets out a scream that feels like sound waves of love that shower over him like cool rain. "Horace my love," she says as they perform an Adimowan type of passionate kiss. Makal turns to the rest of the group. "The Highest is merciful, is he not?" Makal asks. They all nod yes. "You were forgiven. The Highest allowed it to happen. You were fulfilling the prophecy that he spoke of thousands of years ago." I did not even know or understand the reason why, until now.

Angelia stands silent and stunned in front of her mother. Makia, turns all her attention toward her. "You are so beautiful," Makia says. "I died the moment I had to leave you behind. I broke the covenant between Earth and Adimowa, but you paid the price. I knew that you would be taken care of, but I could barely take the pain. Outer darkness was not as painful as the pain I felt when I left you. I wanted very much to stay but I had no choice or say." Makia meets everyone. They all surround her to finally meet the visitor that Dr. Mitchell so often spoke about. They are family now. "You were drawn here for a great purpose, one greater than you could ever have realized," Makal says. "Surely the Highest knows what has happened on Earth already. If the great war has begun again, then Pherous has gone the way of his father Lucipherous, waging war against the Highest." "He has already freed Lucipherous from the pit," Dr. Mitchel says. "He wants it all, Earth, Adimowa, and this new planet that he made." "Then there is no time to waste. I will present you before the council, not for Earth's sake, but Makia's, for it was her love that drew you here." Makal gathers the group, who now shine, their faces are bright, flawless brown and black skin. "We will go to the second level of Adimowa and speak to the council. There is someone there that I want you to meet." Makia smiles. "Where are all our ancestors and loved ones, are they here,?" Louis asks. "I want to see my mother and father." "I do too," Birhan says. "I thought they would be here to welcome us." "We all have someone we would like to see if that's possible?" Dr. Mitchel asks Makal if they can see their dearly departed loved ones. "Is there time for that, he asks? "You are just passing through; that's why they are not here for you. Your ancestors that are here cannot see you or sense your arrival. For one of you, this is not so," Makal says. They all look around wondering what he meant by that. "Look," Makal says to Horace, pointing towards an open area behind him. Dr. Mitchel turns and looks. Waiting for Dr. Mitchel are his many ancestors, all smiling and happy to see him. They have sensed his arrival to Adimowa. Very excited to see him, they all come over to welcome him to Adimowa. The others cannot see what he is seeing. "What is he doing, where is he going?", Louis asks. They are greeting him, Angelia says. "Who", Louis asks. "His ancestors", Fatimah says.

Does this mean, I died?," Dr. Mitchel asks Makal. "The others translated by the way of Enoch but you came the natural way," Makal says "My heart was bad, I must have died in the sarcophagus as we translated," Dr. Mitchel says. "Yes, welcome to Adimowa my son." His ancestors communicate with him all at the same time without saying a word. It is a very spiritual exchange that courses within the family line, like DNA in the blood. Dr. Mitchel communicates a "see you later" to his departed ancestors, letting them know that when this is over they will have a big celebration altogether. The ancestor sense that he has something to do for the Highest before he can rest in peace. His ancestors all disappear back into Adimowa. He walks over to Angelia and grabs her hands. "Looks like I'm here to stay baby", he says "It's ok daddy, Angelia says, "Things are clear now." We will all be together forever If the Highest says the same." They all start to realize that Dr. Mitchel died back in the Great Pyramid, just as everyone else translated. Do you realize my friend that your body will be the only one to be discovered in the Great Pyramid, in the sarcophagus," Birhan gleefully says to Dr. Mitchel. "That's true," Dr. Mitchel says, not really concerned about pyramids anymore. "We must go now, "Makal says. He leads them through the enormous golden doors, onto the street that moves toward the city of Adimowa up into the second Adimowa where the counsel is. In the distance, they can hear the sweet sounds of singing. It's a perfect Adimowan song they have obviously have never heard before. They all understand the words wisping through the air "The Highest is wonderful; our Highest is marvelous", they sing.

Makal stops everyone to tell them something vitally important. "As you enter the city it is important that you all have humility in your heart and believe as a child would. You must remember that the rest of you are but esteemed visitors and guests here. That means that you are still bound to the Earthly covenant." They each stare at Makal, listening to his every instruction very carefully. "Understand this, you must return to the River of Life within twenty-four hours. If you do not, you will be completely changed over and be forced to stay. The laws of Adimowa would take over." The group looks out and down into a large valley made of clear crystal. They see what looks like tall soldiers dressed in glowing burnt orange Adimowan war-like gear. They are getting ready to deploy to a battle. "You see, what you thought you came to reveal, is already known, as you should have known. He wouldn't be the Highest if he couldn't see very far. You are here for a different purpose; you are here to fight my son." Horace stands tall and begins to feel the power of knowing his true purpose. "You came here a different way than the others because of the prophecy. You will no longer be called Horace; you will have a strong Adimowan Arc name. Horusal the warrior of the Highest," Makal says. All the armies of the Highest stand at attention and give a powerful salute and war cry in honor of "Horusal! Horusal!" "All praise belongs to the Highest, Horusal shouts to the armies in Adimowan!" "All praises!" they shout back in the Adimowan tongue.

CHAPTER 14

LIFE IS DEATH

EARTH, VALLEY OF MEGIDDO NORTHERN DISTRICT OF ISRAEL, JULY 19, 2021

Pherous looks to the horizon from the Valley of Megiddo, ancient Jezreel. It is the place prophesied to be the location of the last great battle for Earth. The battle of Armageddon. It is flat and wide as far as the eye can see. A perfect field for war. "Look father," he says as he points up to the sky. They see 400 thousand plus Adimowan warships emerge from multiple fire clouds that descend into the valley. There are so many that they look like a swarm of locusts. Dust and debris fly in the air, as they touchdown. Loyal followers of Pherous and his troops have arrived. With Earth weakened by worldwide nuclear attacks, Pherous begins his final assault on what's left of the world's military. They are no match for his army. The Earth is scorched, nothing but rubble, shadows of what it once was. The Great Pyramids continue to defy time and remain. All who were not killed in the nuclear blasts and radiation now hide in fear of Pherous and his great sweeping army. Wailing and crying are the only things heard in the cities, the strong stench of death fills the air. Lucipherous sits on a type of throne in the old city of Jerusalem, as the ruler of all the Earth, just as he planned it. Pherous sits at the right hand of his father. "Earth is no more my son," Lucipherous says. "We have done it father," Pherous says. "I am the chosen one of my family, chosen to vindicate you, to make war with the Highest and his chosen."

"Yes, you have done well but this is not done yet. They will fight back. We will finish this today," Pherous says. "Then I will take Adimowa in one hour. His little sacred temple will be our place to relieve ourselves." Lucipherous laughs. "I will reveal his secret place, strip him naked in the streets of Adimowa. Then we will cast him down and put him in the pit, Makal and all those who oppose us too." "Down to the ground! Pherous spits on the ground. "I will give you his throne in 3rd Adimowa. I have set you free to rule again!" Lucipherous just listens to his son offer him oblation and reverence. He eats it up. The power feels good to them both, Pherous is in a drunken state of pride, basking in his first victory. He makes a wild proclamation to his fellow fallen brethren of Adimowa. "Gather all the ones made in his image; bring them to me that they might have the privilege and honor of worshiping their new God, Lucipherous. Perhaps I will allow them to be our eternal slaves. To the pit with all those who try to resist." "Do it! "We will not fail you," they say. Pherous' top lieutenants leave his presence to fulfill his proclamation. He makes another decree before his court. "Oh, and it's time for me to wake my chosen ones. I can't believe their President settled for a moon. How sad," he says.

Second Adimowa Eternity
(No Time Zone) Continuous day

Makal leads Dr. Mitchel and the group to the second Adimowa where the council meets. They see all sorts of wondrous things along the way as they walk through Adimowa's golden streets. The people of Adimowa gather and come up to them to greet them and begin to walk behind them, following them. It's as if the crowd knows exactly why they have come to Adimowa. They all arrive at the counsel's chambers in the second Adimowa which is noticeably higher. They enter the counsel's chamber. When Makia enters the chamber, she sees her mother Aita sitting at the head of the counsel's table. Makal addresses his wife. Aita is one of the chief counsel members of the second Adimowa. She sees Angelia and somehow knows that she is family. She stands to her feet with the rest of the council sitting looking on as she lifts her hands in praise to the Highest. Tears of joy fill her eyes, as she says something to Dr. Mitchel in the Adimowan tongue. Aita breaks down as the purest love overtakes her heart like waves of water rolling onto the shore. It swells within her. She makes her way to the family, and they do the same. They are completely lost in the moment. Sammy, Mona, Josh, Zoe, Alfonso, Birhan, Fatimah, and their two little girls Gabriel and Dellina, Louis, and Angelia are all spellbound, as this wonderful family reunion unfolds before them. Adimowa watches and celebrates along with them. Cheers and songs of joy can be heard for miles around; this is a moment written in prophecy coming true. Makal smiles at them all. Makia and Dr. Mitchel turn toward Angelia who is joyfully crying. "Grandma, my Grandmother, you are so beautiful," Angelia says. Aita stretches her arms out to Angelia. Angelia runs to her, falling into her arms. "I knew you would come to me one day. It was written in the great scrolls that you would find your mother through the trail of love left behind, through the secret chamber of the king. You must hurry; your destiny calls to you. The Great War has already begun. Our dear brother Pherous has given himself to the way of his father, Lucipherous. He tried to make war with the Highest thousands of years ago and was cast down, bound in a bottomless pit. His followers were cast into outer darkness. He has recruited their sons and daughters, with those fallen and has made his army strong.

The armies of the Adimowa are ready to stop him," Aita says. "This was written from the beginning," Makal says. He has almost destroyed your Earth. The group looks stunned to hear this, thinking of family and friends left behind. "I'm sorry, but all is not lost," Makia says. "There is a plan for us all and all will make sense to you soon." "How can the destruction of our home and loved ones ever make sense Makia?" Alfonso asks. "Everybody is dead, blown to kingdom come." Alfonso looks at Zoe with very widened eyes as he recalls the strange street preachers in purple and how they said that it would happen just like this. "They said this would happen, they tried to warn everyone, Alfonso says to Zoe. "Yes, I guess so," Zoe says. "There is so much that you do not understand yet," Aita says. "Death is life and life is death, this is a universal truth." "What does that even mean?," Zoe asks somewhat frustrated. "Were you afraid to be born? Did you have any say in that?" Aita asks. "Life is a gift given to you, not a right. Your heart beats without your permission, does it not? It was made so. It is the pride of man that makes him feel that life belongs to him. Is his own. You still see and think as Earthmen do. But this day you will see his glory shine forth and you shall be changed in your understanding of all things." "Why doesn't God do something?" Angelia asks. Horusal looks at the group. "He is moving. He's moving within us right now. When we move, it is him moving through us.

That's what Aita is saying," Horusal explains. Aita smiles in his direction. Dr. Mitchel's mind is illuminated and his understanding of things is far above the rest. They will start to slowly get it in time. "This place is where they all come to," Horusal says.

"It's harvest time." "This time is prophecy being fulfilled, it is called the end of days," Makal says. "The River of Life is full today." At that exact moment, a great trumpet sounds from third Adimowa, igniting the armies of the Highest to mount up for the great war against Pherous and his father's army. Their war cry shakes all of Adimowa to its foundations. Crystal and gold war spaceships power up. The other Arcs of Adimowa prepare their regiments for battle. Makal speaks to the Arcs angels Gabral, Shankal, and Chapal. "You all know what to do. The Highest has issued the order and the trumpet has sounded. This is what we have trained for.

Horusal, the chosen, will coordinate our ground forces." Horusal, formerly known as

Dr. Mitchel, points at himself as if to ask the question, "Who me?" Makia looks at Horusal and smiles, knowing that he is flowing in his true destiny now. Horusal begins to recall all his many years of intense military training and knows that this was always his purpose in life and death. To lead the Armies of the Highest against Pherous in the final battle for it all. He looks over to another part of the room and sees a restored Rodrigo, dressed in a glowing neon blue flight-type suit walking toward Dr. Mitchel and the group. They are overjoyed to see him again. Rodrigo, now known as Rodrigal, greets them all with hugs and kisses. Rodrigo takes Horusal by the hand and leads him away to prepare him for battle. Alfonso, Louis, and Birhan also want to volunteer to fight. But they are denied the chance, for it is not their destiny.

CHAPTER 15

A FAMILY AGAIN

EARTH, VALLEY OF MEGIDDO NORTHERN DISTRICT OF ISRAEL, JULY 21, 2021

B ack on Earth, Pherous looks on the horizon from the valley of Megiddo. He sees 400 thousand plus of his Adimowan warships emerge from the multiple fire clouds that descend from the sky down into the ancient valley of Megiddo. There are so many that they look like a swarm of locusts or bees. "It's beautiful," he says. Dust flies in the air as they touchdown. The loyal followers of Pherous and his troops have arrived. They walk over to Pherous and greet him in the Fallen Adimowan way. With Earth weakened and crippled by worldwide Inter-continental Ballistic missile strikes, Pherous begins his assault on what's left of the world's military.

Some pockets of resistance try to assemble and attack. The world has only one military now. The World's militaries fly overhead in Fighter jets and even do their best to muster up pitiful attacks, trying to inflict some kind of damage to Pherous and his Fallen. They are no match for his armies at all. It's not even really a fair fight at this point.

"This is too easy," Pherous says. "Can anyone challenge me?" he asks. "Who shall bring me down!" he asks in his pride. The Earth is scorched and smoldering, mostly rubble in its major cities, shadows of what they once were. The Great Pyramids, however, continue to defy time and remain in place. All who were not killed in the nuclear blasts and radiation now hide in fear of Pherous and his great sweeping army. Wailing and crying are the only things heard in the cities, the strong stench of death fills the worldwide air. The sun can barely be seen through the smoke and looks red in the sky. They are now living the book of Revelations. Lucipherous sits on a type of throne in the ancient city Jerusalem, the City of David, as the ruler of all the Earth, just as he had planned it. He sits at the right hand of His Father for the moment. He is truly the "Lord of the Flies." Pherous waves his hands and arms back and forth around his face. "I hate this miserable place, especially their flies. They are everywhere. They are the true warriors of this planet. They will feast today on the carcasses of men," Pherous says. "Earth is no more," Lucipherous says. "We have done it," Pherous says. "You have done it," Lucipherous says. Pherous stands up and looks out over his accomplishments. Again in his pride, he says. "I am the chosen one of my family, chosen to vindicate you, to make war with the Highest and his chosen." "Yes, my son, you have done well but this is not done yet. They will fight back," Lucipherous says. "We will finish this today," Pherous says. "You will take Adimowa in one hour. His little sacred temple will be our place to relieve ourselves and revel in what we've done," Pherous says.

Lucipherous laughs. "Yes, I will reveal his secret place to everyone, strip him bare and naked in the golden streets of Adimowa, and show everyone that he is not the Highest but the Lowest. Then we will cast him down and put him in the pit. Makal and all those who oppose us will get it too. Down to the ground you go!

Pherous spits on the ground. I will give you his throne in Third Adimowa and his seat of all power. I have set you free to rule again my father!"Lucipherous just listens to his son offer him oblation and reverence with his grand speech. The power feels good to them both, but Pherous is almost in a completely drunken state of pride, basking in his first victory. He makes a wild proclamation to his fallen brethren that have started to gather around his throne, listening to his strong words. Gather all the humans made in his image, bring them here to me, that they might have the privilege and honor of worshiping their new God, Lucipherous, he says. Perhaps I will have mercy and allow them and be our eternal slaves, Lucipherous says. No Mercy, Pherous shouts! Vaporize all who even try to resist. Go do it, Pherous yells! We will not fail you, they say. Pherous' top lieutenants leave his presence to fulfill his royal proclamation. He makes another decree before his court. Oh, and it's time for me to wake my children going to Callisto. "I can't believe their President settled for a moon. How sad, he says. Wake them," he shouts!

Second Adimowa Eternity
(No Time Zone) Continuous day

Samuel and Joshua stand ready to fight. "You all know what to do, the Highest has issued the order and the trumpet has sounded, Makal says They all let out the loudest roar you could ever imagine. It shakes all of Adimowa. This is what we have trained for, this is why we were born into the rivers of Adimowa. We will send Pherous and his rejects straight back into the darkness they came from. We will show them what happens when you mock and scorn the Highest, what happens when you try to fight his Armies. We will be his hands today, we will be one stick in his hands. We will be his word in motion and we will shout in the victory that will shake the universe. Each of you has known the prophecies and it stands before us today. Makal points straight at Horusal. They all shout out for joy over the fact that Horusal, the chosen of Adimowa and Earth, has finally come to join them in battle. We will not fail. Horusal, will coordinate our ground forces and he will be the lead. Horusal, formally known as Dr. Mitchel, points at himself as if to ask the question, "who me?" Makia looks over at Horusal and proudly smiles, knowing that he is flowing in his true destiny now. Horusal begins to recall all of his many years of intense earthly military training and knows that this was always his purpose. To lead the Armies of the Highest against Pherous in the final battle for it all.

He looks over to another part of the room and sees a restored Rodrigo, dressed in a white flight-type suit walking towards him and the group. They are overjoyed to see him again. Rodrigo, now known as Rodrigal, greets them all with hugs. "I'm so happy to see," Horusal says. "You as well my friend, Rodrigal says. I am whole again," he says. "I can see that," Horusal says, with a big smile on his face. "I see that you've crossed over too," Rodrigal says. "Yes I did and fine with that, Horusal says. I died on my way here, my body is in the sarcophagus in the King's Chamber." "So you didn't need that stone after all, oh well," Rodrigal chuckles. "Guess not, Horusal says. "Will you fight beside me today," Horusal asks? "Not only will I fight beside you, but if you give me the honor today, I will be your armor-bearer," Rodrigal says with great pride. The Armies of Adimowa start to light up their halos and give the Highest shouts of praise and adoration as they prepare themselves for battle. Rodrigal takes Horusal and leads him away to prepare him for battle. Alfonso, Louis, and Birhan also want to volunteer to fight. But they are denied the chance, for it is not their destiny. To everyone's surprise, Sammy and Joshua are allowed to go and fight. Angelia goes over to her boys who are at this moment completely grown men. She hugs them both. "I trust in the Highest now, you do the same and come back to us." "We will Mother," they both say. Louis looks on and smiles at them proudly. "Trust and Honor," he says. "I want to fight too," Mona says to Aita. "You will my dear," Aita says. Mona looks up and smiles back at Aita. She holds up Mona's chin in her hands.

Horusal stands in front of Rodrigal with both his arms out to his side. Rodrigal has all of Horusal's new battle gear and clothing. Rodrigal lays the orange glowing war garment over Horusals shoulders, the garment does the rest. It climbs onto him and spreads over his body. That tickles, Horusals says. "This battle clothing is yours and yours alone. It knows you and has become a part of you. It and you will be tried and tested during the battle. It will move with you, my friend. When you speak it will speak for you," Rodrigal says. "What does that mean," Horusal asks. "Point your right arm, hand, and fingers over there," Rodrigal says. Horusal points his arm and holds his right forearm with his left. Now think and speak a trigger word as if Pherous was standing right there. Horusal's face changes to reflect his rage as he says the word Peace. His voice waves of energy flash out of his fingers like a laser beam of lightning, completely destroying the area where it strikes. Horusal looks at his hand in disbelief at the power that flows through him. "You are ready," Rodigal says. Horusal looks like a comic book hero standing there in his glowing Adimowan war suit. We are ready, thank you brother, Horusal says. No thank you for selecting me back then and choosing me to help you.

CHAPTER 16

THE CHILDREN OF THE FALLEN

EARTH, VALLEY OF MEGIDDO NORTHERN DISTRICT OF ISRAEL, JULY 22, 2021 3:26 (IST)

T he vast army of Pherous' fallen rejects of Adimowa, stands in battle array at the rim of the Valley of Megiddo, itching to be released upon the Earth. The army lets out a thunderous shout in the Adimowan tongue. It sounds like actual thunder clapping that shakes the very ground beneath them. Pherous levitates to hover over them. "I am Pherous, son of Lucipherous, son of Gethera! This is the night that we have longed for and were created for. This night, we shall bring down the one who brought your fathers to nothing and cast them into outer darkness forever. The one that wounded you, made you orphans and made you the children of the fallen. This night, we will fight him, for he comes. He pauses and looks to the left of him. Behold, the one called Lucipherous!" They all begin to scream and shout as Lucipherous steps up into their view. He is strong and powerful looking, a sight to behold and fear.

The crowd of fallen Adimowans all rise to their feet and roar. They howl, they scream and shout their war chants all in honor of Lucipherous, the original rebel in their eyes. If they had any doubts about complete victory, all those have now gone away. Pherous lifts his arms and shakes his fists with both hands. His pilots run to their ships; his soldiers come to attention in their ranks. Pherous looks to the sky and smiles. I want it all, he says to himself. Pherous starts to laugh uncontrollably, still looking to the sky.

"Wake my children up," he shouts, referring to the President's chosen heading to Callisto! "It's time to awaken my children and begin my new earth, that I will rule with an iron fist before my father." Pherous' troops bring in several of the humans that they have locked in electromagnetic cages like animals. "Yes, bring them here. Let them out." They take the earthmen out of the cage and make them stand before Pherous. He addresses the first one in front of him. He points to Lucipherous and says, "Worship him." The man that stands in front of him says no, and begins to shout " I bind you Satan, I bind you in the name of Jesus!" "Greek, a very privative language, I love that," Pherous says.

The preacher man has his eyes closed tight, with hands lifted up. No, bind you," Pherous says, in an angry and annoyed voice. Pherous grabs his right forearm and shouts a very sharp voice wave at the preacher that cuts his head clean off, with no blood at all. Completely cauterized by the hot waves. The head rolls down and comes to a stop at the feet of Lucipherous. The rest of the prisoners fall to their knees and begin to worship and adore Lucipherous, at least in appearance. Pherous instructs his royal guard to allow the worshippers to live as slaves to the court as a reward for their obedience. "Very good he says, very good, now let's go and finish this!," Pherous shouts.

CHAPTER 17

CALLISTO IS YOURS

SHIP TO CALLISTO INTERSTELLAR SPACE JULY 23, 021

P herous, we are here in Callisto, your children have arrived safely. We only lost a few on the way. Callisto is yours, he says!" Each of the two Adimowan pilots in the ships simultaneously receives a message from Pherous through their ship's console headset. They push a button to start the process to reverse the cryogenic sleep process. They wake up fairly quickly and start to stir around the ship as they rise. The President of the United States of Callisto is one of the first to wake up takes this opportunity to address the chosen as they all slowly regain consciousness. "Good morning to you all. I trust that you can all hear me in each of the ships and pods, I hope that you can. I know that many of you are still trying to gather yourselves and the last thing that you want to hear is my voice so please forgive me ahead of time. What a glorious day this truly will be my brothers and sisters. I can barely wrap my head around this. It is so glorious and magnificent.

We are making new history for mankind today. We literally have boldly gone where no man has ever gone. We have made it safely to our new home. As you can all see out of your cabin windows it is everything that we could have ever hoped it would be, and more. It is the Garden of Eden all over again. I am overjoyed to say, welcome to Callisto!" All the ship's passengers explode with joy and exuberant celebration.

The advance team on Callisto, made up of ten scientists, and five engineers, have gathered at the runway to welcome the colony of ships as they make their historic arrival. There are several of the Adimowan fallen there to greet them as well. Just as the ships are about 300 feet from the surface of Callisto, each Adimowan pilot pushes a button on their console. As soon as the button is engaged it seals up the cabin area where the Adimowan pilots are. Then it causes the cabin to break away from the rest of the ship and jettison away. The President and the others gently float to the ground to the surface for a perfect landing. "The Eagle has landed again," the President says over the intercom. They all start to clap and woot. "We did it," Jeff says to his team. After a few minutes of adjustments and cool down, the doors to all the ships open simultaneously. The chosen members of Callisto and their family members begin to walk from the ships.

The President and his family are the first to step foot on the new planet. A full band plays "Hail to the Chief." The G7 members, then the powerful families come out. Chairman Roth and his family make their triumphant exit followed by the other secretaries and heads of state. It is all very ceremonious. Some of the fallen Adimowans step forward to greet the President and the others as they disembark. "You smell that," the President says to his wife and kids. "What," Gladys asks. "Clean air. It is so beautiful." "Yes, it is Randolph, Gladys says with a big smile on her face. "I can't believe we did it," The President says.

The kids look up at the large black Adimowans and then run over to their father. "Daddy, they are black," his daughter, Sophia says. "No situation can be perfect baby, try not to stare at them, he says" His wife Gladys looks stunned to see that the Adimowans all seem to be black looking. Mark and Gregory look at the aliens with sour faces and just keep walking. Two of the Adimowan delegates walk over to the President and his family and one bends down. "Welcome to Callisto, I greet you in the name of Prince Pherous," he says. "Thank you," says the President, shaking his enormous hand. One by one the members of the advance team greet the President.

Jeff Aragon and his team offload their gear and equipment. "Keep a close watch on our stuff and make sure nobody goes in those duffle bags," Jeff says to one of his men. The President, Jeff, and the entire chosen load up in busses and SUVs and are taken away to the main city and housing area to get settled in. Jeff sits in the passenger's side of the black SUV and marvels at the tall Adimowan delegates that are still standing near the ships. "This shit is unbelievable," Jeff says to the driver. "We are sitting here in an SUV on a moon, that looks like the Earth, that's been destroyed. All made possible by the help of some big-ass oversized Labron James looking like aliens all around us to keep us in check. Why? Anybody ever takes the time to ask why?" I guess I would rather be here than dead. "You talking to me?", the driver asks Jeff. "Nah, just forget it," Jeff says, let's roll." The rather large convoy of vehicles speeds off on the freshly paved highway with absolutely zero traffic.

CHAPTER 18

THE GREAT BATTLE

THE VALLEY OF MEGIDDO NORTHERN DISTRICT OF ISRAEL, JULY 24, 2021 6:43 PM (IST)

Pherous hovers on a cloud as he speaks to his sea of the fallen of Adimowa. "Callisto is ours my brothers!", Pherous shouts. His ground forces erupt with vigorous cheers, excited over their second campaign victory. One part of Pherous' army stays in the camp and moves out to their various fighting positions. They dig into their own ready fighting positions, in the Adimowan warfare way, waiting for Adimowa's imminent attack. Their warships sit ready, waiting for the first wave of attacks to come. Another portion of Pherous' air force and army continually search the Earth to kill or capture any humans left on the planet. "Pherous, why do we wait here for Adimowa to attack us? Why don't we attack them first?", one of his officers asks. Pherous looks at him with a rotten and disgusted look "Why? Because he is coming to me, that's why.

My father is free now. He will wage the war in Adimowa while I lay waste to this place. They know what's coming, there is no element of surprise here." "He already knows what is going on what we are planning. We can only give him what he knows. The warrior looks confused, not really understanding what he said. "He is coming to us with his armies, and we will be ready for him this time. May I ask another question, great Pherous? the warrior cautiously asks. Pherous nods "yes" to him. "Can we win this war? he asks. "He can be beaten through division in his ranks and doubt placed in his warriors' hearts. We must split them in heart and will and then and only then will we be able to split the prophecy. We will lure them in close and then destroy all of them.

My father made that mistake, taking the fight straight to the Highest in Adimowa. Remember, I've just succeeded in exterminating man, all of his little beloved creations, made in his image. "I still don't understand why he loves his people so much," Pherous says. They are almost all but dead now, sent back to the River, but no longer on his Earth. And Callisto, well that is my mine, my creation and he will hate that too, or at least I hope so. I have his so-called wicked ones there now, colonizing the new world as we speak." The warrior's eyes light up with new understanding. "We will be victorious over him, over all of them!" the warrior shouts! Pherous looks at him and smiles, knowing that he has strengthened the warrior's resolve with his words.

First Adimowa, Eternity
(No Time Zone), Continuous daylight

Horusal and Rodrigal walk back to where the armies of Adimowa are staging and strategizing for the coming attack. They look back and wave at the rest of the group as they climb aboard their ships. Horusal is the pilot and Rodrigo is his co-pilot. Makal and the Arcs signal to the mighty Adimowan warriors that it is time to launch the offensive against Pherous and Lucipherous' armies that have all but destroy the Earth. The Arcs, Makal, Gabral, Chapal, Shankal, Rodrigal, and Horusal take the lead. Thousands of Adimowan ships fill the skies of Adimowa. They all hover in vast formations across the sky, as far as the eye can see. They all start to generate a powerful and very large fiery cloud around them all. Then, in a flash, all of their ships hyperjump at translation light speed to the gateway into planet Earth.

Former United States Airforce Facility,
Area 51, (Sector 1)
Nevada "The Extraterrestrial Highway",
July 24, 2021, 9:19 PM (PTS)

Most of the organized forces of Adimowa's army emerge from a fire cloud over Nevada's, near Area 51 near what was called the extraterrestrial highway. The former secret Air Force Facility is completely destroyed and completely laid to waste. Burnt and bombed-out tanks and airplanes lay about the desert floor. The fire from the cloud falls from the interstellar war crafts as they all form up and speed from there to the Valley of Megiddo. One by one they jump to hyper speed and leave. "Can you believe that we are doing this?" Rodrigal says. " I know, but I can't think of any other place that I would rather be right now. This is our time to be used by the Highest, we are that smooth stone that David used to kill Goliath, " Horusal says. They quickly arrive in the Valley of Megiddo. We're here, here we go my brother, Horusal says. Makal signals that they have arrived. They all begin to make their entry back into the atmosphere. "Sound the trumpets," Makal shouts. Upon receiving Makal's, instructions every Adimowan craft sounds its onboard horn. The Sound waves tear through the sky.

CHAPTER 19
ALL HEAVEN BREAKS LOOSE
EARTH, VALLEY OF MEGIDDO NORTHERN DISTRICT OF ISRAEL, JULY 27, 2021 8:43 PM (IST)

Pherous and his army wait for the imminent Adimowan attack. They carefully watch the skies as their anxiety builds within them. As they watch and wait, suddenly a great trumpet sound is heard above them and all around them. Pherous' army shouts a war cry back at the sky and then to his troops. Thunder and lightning crack, as the clouds roll back like giant scrolls. Complete silence in the rest of the world that is still licking its wounds. Then all Adimowa breaks loose from fiery clouds in the sky. "Here they come,!" Pherous yells. Thousands of thousands clash in battle. The Great War, the battle of Armageddon is underway.

Pherous uses his voice waves to tear through the soldiers of Adimowa. His voice bores straight through many of them while he deflects attacks that try to end his eternal life. The battle rages in the air and on the ground. Pherous fights with his army in the air. The weapon of choice for both sides is the voice. Their voices come out like sharp waves of knives, able to cut to the spirit and soul. Horusal and Rodrigal takedown about fifteen of Pherous' ships with their voices, while the other Arcs fight valiantly doing the same. They are learning as they go, becoming more and more proficient with the new Adimowan weapons they now possess. Horusal locks in with a very fierce-looking fallen Adimowan soldier, He shouts toward him while grabbing his right forearm. A beam of voice wave energy comes out of his four right fingers. He is no match for Horusal, He quickly takes him out, and it's on to the next one. The back of Horusal's head begins to shine bright as the sun. He starts to feel the strength of the

Highest with him. Joshua sends out his voice and smashes through a brick wall, then he ends several of the fallen. Sammy runs and jumps straight into the chest of a warrior. He looks like David from the bible when he fought Goliath.

The sky looks like a southern evening with fireflies in the air. On the ground, Horusal and Chapal lead forces in a fierce and violent onslaught. Beams of voice projectiles zip back and forth. Pherous' fallen explode when they are hit with a voice wave. Physical sound waves, not bullets, fill the skies and ground. This is the most epic battle that has ever been known in the history of ancient and modern warfare. It is the prophecy being fulfilled. "We have them now!" Pherous yells. Pherous' forces seem to be winning the physical and spiritual fight, ending many of the Adimowan warriors. "Get that regiment over to that ridge now!" Horusal yells to Rodrigal. "We can cut them off there now!"

Rodrigal gets the message out to the rest of the troops. The regiment moves quickly to the ridge and cuts Pherous' troops off, in a surprise attack. "Behind us!" one of Pherous' warriors yells. Pherous and the others turn from the battle in front of them and begin to fight the massive attack from the rear. Pherous and his troops sustain many losses and decide to retreat from the battle. After ordering his armies to advance and sending them to their demise, Pherous escapes through the side. Makal sees him slithering away like the coward he is and keeps his eyes on him as he tries to escape. Alone, Makal valiantly chases him. "You are all mine," Makal says to himself going in pursuit of Pherous.

CHAPTER 20
VENGEANCE IS MINE
FIRST ADIMOWA, ETERNITY (NO TIME ZONE), CONTINUOUS DAYLIGHT

It's another almost indescribably beautiful day in the Kingdom of Adimowa. "I am amazed at how much Adimowa looks like Earth but so much better in every way," Birhan says. "I almost could stay here. It's good to know this is the end game," Louis says. An alarm begins to sound throughout all Adimowa. The first Adimowa is under full attack by Lucipherous and his army of the fallen. They strike from every side of Adimowa at the same time. The citizens of Adimowa run for cover while the small forces left behind call to arms. Louis, Alfonso Nazario, and Birhan gather the rest of the group and run to Makia and Aita. First

Adimowa is being ripped apart by Lucipherous. The people of Adimowa run in terror screaming and they are ended by the thousands. "It is Lucipherous, he has returned!," an Adimowan says in terror. How did he escape from the pit?" Aita asks. "Get the children to Third Adimowa, he will not have mercy on anyone? Aita says to Angelia and Zoe.

You should go too," says Aita to Makia. "He will try to end us all spiritually, from which there is no return." "No mother, we must all fight him now, or he will take Adimowa!" Makia says! Angelia, Zoe, and Fatimah take the children higher to Third Adimowa while the others turn to First Adimowa to help in the fight against Lucipherous. Lucipherous and his fallen army engage what's left of Makal's army in a battle as fierce as the one that rages on Earth. Everyone in Adimowa fights. Makia and the others fight like seasoned warriors. The great war now rages on two fronts. They have been successful at splitting the prophecy, just as Pherous had planned but will that be enough to secure him the victory.

Earth, Valley of Megiddo
Northern District of Israel
July 31, 2021, 7:57 PM (IST)

Pherous' army is almost completely defeated. Oh, how the fallen have fallen mightily. Makal, his other Arcs, Samuel and Joshua have ended most of Pherous' warriors. Pherous is sitting atop a pile of rubble as a makeshift throne in the city of David, in Jerusalem, waiting for his fate to arrive. Judgment for his high crimes. He knows that all is lost for him, or is it? Makal arrives where Pherous sits waiting for his end to come. Horusal, Rodrigal, and the other Arcs are still finishing up small pockets of resistance in the valley. Samuel and Joshua continue to tirelessly fight. Pherous is finally cornered, with nowhere else to run or hide. "Pherous, it's over for you and your fallen ones," Makal says. "Why do you follow in your father's footsteps?" he asks. "I and my father are one," Pherous says. "I respect you Makal." "You respect nothing Pherous," Makal says. You should not have come here Makal; you are not strong enough for me. Makal circles him. "I put your father in his place long ago and now I will put you in yours!" Makal moves into his fighting stance and then rushes over to fight Pherous. Pherous, roars and transforms into his beast form. He takes on his monstrous animal form.

Pherous rushes from where he sits and crouches like a young lion. Makal looks surprised to see Pherous in this ancient form. "A beast on all fours suits you," Makal says. Pherous and Makal clash in a battle of the gods. They all but destroy what's left of Jerusalem. "I told you that you are no match for me!" Pherous yells. "You are too weak to defeat me Arc! It is a new day and Adimowa will be ours no matter what you do to me! Pherous finally overcomes Makal and gets him in a chokehold. I'm going to end you now

Makal, but first, you should know that even now my father Lucipherous is free. He makes war with the Highest now; he has taken Adimowa. That was the plan. I'm sure your family is dead too." Pherous lets out a loud cry and then takes his arm and slashes Makal's throat with his voice waves. Light comes from the gaping slit in Makal's huge neck. His light behind his head goes to a flicker and then goes out, completely dark. Makal is ended spiritually forever. Pherous lets Makal's lightless body fall to the ground. Makal turns white from head to toe.

Pherous lets out a loud roar that can be heard for miles. At that moment Horusal, Rodrigal, Samuel, and Joshua arrive to see Pherous standing over the body of Makal. "No!" they all yell. "Yes," Pherous says as he turns toward them. He bows to him. "Chosen one of the prophecies. It was you all the time," Pherous says. He looks at Rodrigal. "I remember you, I split you in two." He laughs at the rhyme he makes. "I got this one,"

Horusal says. "You've got this one," Pherous laughs. "End him Pawpaw," Sammy says. Joshua pounds his chest as a salute to his brave grandpa. Rodrigal looks at Pherous then nods at Horusal and runs away. "Ok, let's play," Horusal says. He circles around Pherous still in his animal form. Horusal has Adimowan special weapons to fight. "I shall send you to outer darkness by the help of the Highest."

Pherous hears his voice wave weapon, knowing that he can do just what he says. "You've really changed since the last time I saw you, so angry and focused. I like the changes." Pherous grabs his right forearm and then runs to attack Horusal, shouting his voice projectiles. They fight until the absolute bitter end, completely destroying the surrounding area. Horusal now has strength and skill beyond his own belief. Horusal's words are powerful and accurate. At that moment the other Arcs arrive with Samuel and Joshua and several other Adimowan warriors. Rodrigal has brought them to see this great fight. It goes on and on for what seems like hours. It is actually beautiful to see. They move like dancers in a ballet, poetry in motion as they fight. Pherous says something to Horusal while in a hold. "I will be sad to end you Horace." "That's Horusal," he says. "You've really grown strong," Pherous says. "Maybe you can be my slave if you stop now. Or maybe my father's pet." "My voice will separate you from the highest forever," Horusal says. Pherous appears somewhat tired from the fight. Voice waves fly sharply through the air. "My father has taken Adimowa, you cannot stop us chosen one. He is defeated so you fight for nothing."

Horusal knows that he must end this now in order to get back to Adimowa and stop Lucipherous. Horusal makes his critical move. He leaps at Pherous and flips over his head. Horusal lands on his back and lands a deadly blow to his neck, in the word "Peace", sharply passing through his hand. Pherous lets out a roar and falls to his knees. Horusal screams out a loud war cry. The onlookers go up in cheers. Pherous hangs on for a few more moments to say "Earthman, tell Makia she made a wise choice." "Damn right she did," Horusal says. "Father!" Pherous cries out and then falls flat onto his face.

Horusal rolls him over. He takes his living knife and removes Pherous' partially glowing halo gem from the back of his wooly head and holds it high in the air as the light goes out. Horusal lets out a loud war cry that everyone left on the battlefield hears. Pherous turns back into his Adimowan form. He is dead, physically and spiritually." Lights out," Horusal says. The great battle for Earth has ended with the eternal death of Pherous. Horusal turns to the other Arcs. There is no time to celebrate the victory for they know that they must leave right now if they are to fight Lucipherous and prevent him from destroying all of Adimowa. "We will take him home," Horusal says. They all stop and carry the body of Mikal back to the ships. They quickly turn to their ships and return to Nevada area 51 where they will all re-enter the portal and make their exit back to Adimowa. In a flash, faster than the speed of light, they all jump from the valley of Megiddo back to Nevada in Area 51. It only takes a few moments for them to generate the fiery cloud to get back to Adimowa. What's left of the armies of Adimowa rides the fiery cloud back to make war with Lucipherous.

Second Adimowa Eternity
(No Time Zone) Continuous daylight

Adimowan warriors left behind to guard Adimowa, fight hard to hold back Lucipherous and his fallen miscreants. "Run, you cowards," Lucipherous yells, chasing and ending all who he encounters. He has already taken the first Adimowa and has pushed those left to fight up to Second Adimowa. The first Adimowa burns with no one to extinguish it. The bodies fill the streets of gold. Makia and the others fight alongside the warriors. Louis and Alfonso fight like seasoned warriors and so does Birhan. "Lucipherous has made his way into the great temple on second Adimowa," Makia yells to the others. "Where are you?" Lucipherous shouts to the Highest while standing in his sacred temple. He shouts his voice waves out in indiscriminately from his right hand, hoping to hit anyone or anything. He is hell-bent on seeing the complete destruction of all of Adimowa and every Adimowan. They have fought hard and ended many of Lucipherous' fallen but they have proven to be too strong to be stopped.

They all retreat deep into second Adimowa and look for a place to hide. They begin to call out to the Highest. Lucipherous destroys everything that he comes in contact with. "We cannot hold him off for much longer," Aita says. "They grow stronger and stronger with every wave of attack." Just then Lucipherous and his rebels almost overtake all of second Adimowa. As they advance to land their death blow, they all hear the sound of a great trumpet blast. Everyone, including Lucipherous, stops right in their tracks and looks up. "Look," they all say, pointing to the sky. The skies of Adimowa open and are filled with what's left of the Adimowan warriors from the great battle on Earth. Reinforcements have finally arrived. It's Horusal, Rodrigal, Samuel, Joshua, and the other Arcs just in time to fight Lucipherous and stop him from taking all of Adimowa. "Thank the Highest they have arrived," says Aita, exhausted from the battle.

"Come!" Lucipherous shouts at them. "Come and take me if you can, you son's of motherless dogs!" Another fierce battle erupts with voice waves flying all around. Joshua is struck by a voice wave from an Adimowan fallen warrior that sends him flying across the room and into a burning bush. Samuel stops the fight to check on his brother. He runs over to him and rolls him over to put the fire out. "You ok," Samuel asks. "Yes, I'm fine little brother Joshua says, It was only my shoulder." The voice wave only grazed him just enough for him to lose his balance and footing. "Well get up and fight and I ain't little no more," Sammy says. "I'm still in the fight," Joshua says, as he springs to his feet. At that moment a great light shone from Third Adimowa. Everyone stops in their tracks, including Lucipherous. "Finally, he comes," Lucipherous says. "It is the Adim, Son of the Highest, come to fight with Lucipherous one last time, says Aita, looking to the Adimowan sky with hope in her eyes. He floats down in the middle of everyone. He is taller than everyone else and has a face of dark brass, with light surrounding his entire wooly head. He shines bright as a morning star. His voice is powerful as a raging river. "I am The King of Adimowa an Earth, he says with all power and authority. "Where is your father boy," Lucipherous asks, trying very hard to show great disrespect to Adim. " If you see me, you've seen him Adim replies. "Your son is dead Lucipherous, because of you," Adim says. "No!," Lucipherous shouts with rage, as he rushes to attack the Adim with his voice waves. Adim just brushes off his sharp voice attack then grabs Lucipherous with little to no effort at all. Adim subdues and stretches Lucipherous high above his head, then quickly breaks his back. The rest of Adimowa sees Adim locked in battle and since the victory. They crush what's left of Lucipherous rebellion.

The Adim throws Lucipherous' almost lightless body down to the golden ground. Lucipherous is in excruciating pain, on his knees before the King of Kings, his darkened halo begins to finally shine brighter as he is humbled. The Adim kicks Lucipherous in the chest with his massive foot, sending him backward. "Are you going to send me to the pit again? "You are so weak, Lucipherous says with disdain. That is the real reason I had to rebel." "You mistake my patience, for weakness but this day you shall be judged for your crimes against the Highest," Adim says. Adim closes his eyes and lifts his arms towards the sky. Lucipherous is levitated and lifted up from where he laid on the golden ground. By the power of Adim's thoughts, Lucipherous slowly rises into the air in front of everyone present. Adim's mouth opens, and with a shout to end all shouts, his powerful voice waves in the words "you chose Darkness!," rips Lucipherous into several thick pieces of ash that fall to the ground. Instantly, a mighty wind from third Adimowa, like the breath of the Highest, sweeps through all of Adimowa, taking away the lightless bodies of every fallen warrior including Lucipherous. "There will never be war in Adimowa again," declares Adim. All the people of Adimowa let out a tremendous victory cry, such as was never heard by men or Adimowans.

Songs of praise break out and rise throughout the kingdom. Makia looks for Horusal and Makal. He looks for her too. "Makia," he says. She turns to him and sees the look in his eyes. "Where is my father?" she asks. He looks down. "It's ok," she says. Aita stands behind her. "We shall mourn eight days for him, the Adimowan way." "Where is he?"

Aita asks. Rodrigal takes her by the hand and takes her away to see his body. Horusal takes Makia in his arms and embraces her tight. Birhan hugs Fatimah, Louis holds Angelia, Alfonso hugs Zoe, the boys hug the girls, laughing and dancing around. Adim has restored order to all of Adimowa. They gather at the Adim's feet, like children around their father; then he tells them to rise. "There is much you still do not understand and there is much for you to learn," Adim says. "Horusal, Makia, you have done all, and done well. Earth is no more, it was destroyed, removed by man's greed, lust, and thirst for knowledge. The righteous judgment of the Highest has gone forth." "A great harvest of souls has come to the River of Life to Adimowa this day.

These souls have been reborn into Adimowa to take the place of the fallen ones that chose eternal death in darkness over eternal life in the light. Horusal, son of man, made in his image, in His likeness, he has spoken. Man shall survive, man, shall live on because his spirit lives on in you. The very same breath of life given to man on the first day is inside of you now, the same breath that was in the first Adim and Evi. It cannot die, for it is part of the Great Spirit, his spirit in you even now. Horusal, your love for Makia, and the faith you never knew you had in the Highest brought you here to find her and to defend Adimowa from its enemies. You and the others shall go to Callisto and restore, rebuild and replenish the new Earth, man's creation, also made in his image. So let it be," Adim says. Everyone tries to look at the light but cannot because he is light. Songs of worship and adoration fill the air all over Adimowa. Everything fades to white in perfect "peace" that cannot be understood or fully comprehended.

CHAPTER 21

HELL TO PAY

CALLISTO, SECOND LARGEST MOON OF JUPITAR, JULY 31, 2021 (SPACE TIME)

The chosen of Callisto frolic and play without a single care in this new world. The President sits with his friends and family enjoying the peace that Pherous has created for them. It's a perfect day, the purest blue skies, fresh air, green and lush, different fruit trees never seen before. Blue crystal-clear rivers, lakes, and oceans. New animals and old, together. Their children run and play old familiar games like "hide and go seek". Some of the Adimowan fallen stand around and even play. "I can't stop thinking about all those people who died Randolph," Gladys says. "People we have known for years, just gone and forgotten." "I'm going to say something right now that I know you will not like but who gives a damn about that now. Look at all this," the President says. "You are so hard and cold, I don't even recognize you anymore," Gladys says. The President now realizes the gift in the seed that Pherous placed in him when they first met at Area 51. The seed of complete coldness that can cut anyone loose and never even look back. "I literally lasso the moon and give it to you and still you complain," he says. "This is not right. It won't last, it can't," Gladys says. "You're such a negative Nancy sometimes, you know that," The President says. "We will have hell to pay for this one whether you believe it or not Mr. President," says Gladys. "You're welcome!" he shouts. "Kids come here," he says. All three of his children stop kicking the soccer ball and come over to him. "Yes, father," they all say, wondering what he wants. The President just hugs them in a big bear hug. "I just wanted a hug all of you and tell you that I love me." They all laugh at his "joke" and hug him tightly, then run back to play. "You see them, he asks, they are why we did this. Nothing is more important, no one is." "I love me, Randolph," she sarcastically asks, really." Gladys walks away and heads back to their futuristic-looking, luxurious, dome-shaped mansion. "I was joking," the President says. Be ready for the big celebration tonight!" "Such a fucking hypocrite," he mutters under his breath. "Enjoy the VIP treatment you receive because of me tonight." He continues to just relax in his presidential lawn chair and finishes his Tom Collins cocktail, then signals for a refill. Gladys looks back at Randolph and just shakes her head. At the same time,

Jeff Aragon and his team are laid up in a jacuzzi with naked Callisto women. There are no laws or restrictions here, anything goes. "We made it, this is like heaven man," Jeff says to his boys. "I can't believe that they let us come too", says a team member. "We put in the work, and this is the reward fellas. Enjoy it. Ladies, do me a favor, can you give us a minute?" They all stand up, climb out and leave completely naked. "Come back in exactly ten minutes, ok?" They all nod yes. "Damn, they fine," Jeff says. "We could run this place Jeff, who would stop us?" "Don't even think about it dog, Pherous runs this shit. This is "next level". That is one big black dude, he looks like he could rip your head off." Jeff has a curious look on his face. "Yeah, I guess so," Jeff says, thinking of a possible new mission. "Plus, we don't have any weapons or gear here." "You sure about that?" Jeff says. They all look at him with widened eyes. "I never get into a situation that I can't get out of." They all laugh as they consider the idea of taking over Callisto. "Maybe I'll change the mission tonight at the big party. They should have never let us come here if they wanted it to be perfect." The girls return to the jacuzzi with drinks and nothing else. "How can you think about anything else right now, look at these women," says a team member? "Welcome back ladies," Jeff says, now where were we?" he asks. The girls hand off the drinks to the men and then sink back into the bubbling hot water.

CHAPTER 22
WRITING ON THE WALL
CALLISTO, THE SECOND LARGEST MOON OF JUPITAR, JULY 31, 2021 (SPACE TIME)

The President's hand-selected of Callisto all sit down for an evening meal which is the most fantastic feast with all the trimmings. It is a spread prepared for a king and his court. It is a very formal black-tie event for only the upper crust of the chosen. Men and women are dressed to impress the other wealthy that are all present. They have already figured out that even in paradise you need class levels and servants. They have made the lower rich serve the higher. Someone's got to do it. "Hurry up, wow they are slow," the former Secretary of the Department of Homeland Security Thomas Mann says to his wife Margarette. "Still working the kinks out I guess Thomas," she says. "Yes, I guess so he says. Among the lower rich, there are celebrity chefs and fancy restaurant owners that were imported just to serve. They even have those just for their high levels of entertainment like that of JayZ, Beyoncé, Taylor Swift, or Kanye West, to name a few. Only the best for the best.

They perform three shows a day for a captive audience. It was the price of a ticket to the new world. Even on a faraway moon, the laws of the Highest still apply. Work by the sweat of your brow, until you die. There is still the struggle between the haves and the have nots. "Let us take a moment to remember the lives that were lost on earth and let us now give thanks for escaping safely. We are so blessed to be chosen. We thank Lord Pherous for all that he has done for each of us," the President says. "They have no idea that Pherous and his father are both dead and their rebellions on Earth and in Adimowa have been completely crushed. "I know that Pherous is not here right now, but can we just give him a standing ovation?" They all spring to their feet and vigorously clap toward the sky. They are so grateful that some even start to weep.

Then, At the rear of the banquet hall enters Jeff Aragon and his boys from Ghost Recon through the large decorative wooden doors. The doors fly open making a loud and disruptive noise. Every head turns to see the commotion. Jeff struts in like a rooster that's really a fox in the hen house. "Praise da "Lawd" Pherous!" Jeff shouts, in an old southern accent. Everyone's eyes are fixed on Jeff, the man in charge. Jeff has a large green duffel bag with several AR-15 assault rifles in it, ready to go. They have a plan to take control and kill whoever gets in their way. Jeff and the other team members reach inside the bag and pull out their weapons. Everyone in the room is stunned. "Surprise you rich assholes," Jeff says. Jeff and the men were raised mostly poor and from small towns in the South and mid-West. "Everybody stay calm," Jeff shouts so everyone in the big room can hear him. The rest of the team spreads out and takes their sector to cover. "Looks like there's trouble in paradise," one of his team says to a couple seated at a table.

The President tries to console his wife Gladys and then goes up to the raised center table mic. "Just what do you think you're doing, this is not the place for this," he nervously says, trying to show that he is still in control somehow. Jeff and two of his miscreants start to walk down the center aisle between the tables, toward the center table where the President is. All eyes are fixed on them, waiting to see what they do next. Some people fall to the ground and hope that they don't get shot. "Everybody stay seated at your tables with your palms facing down on the table, Jeff orders. "Just relax." "There's been a change of plans for Callisto and it starts right now," he says. I'm the new President and these guys are my cabinet members." We no longer take orders from you Randolph," Jeff says sarcastically. The people hiding under the tables start to slowly come back up and do what he says. "You see all this, Jeff asks? All of this is ours now," Jeff says. One of the President's men tries to be brave and thinks he can catch one of Jeff's men off guard and maybe take his weapon away and possibly save the day. He doesn't get very far and is immediately shot several times by one of Jeff's bloodthirsty team members.

The body is ripped to shreds by the powerful assault rifle rounds. Everyone in the party screams and is mortified by what they just have witnessed. Most begin to wail and cry in terror and disbelief. "Now see, I told you all to stay calm and nobody would have to die today, but there you go. Now, who wants to be next?" Jeff asks, looking at the rest of the President's guards. "Anybody else wants to earn a posthumous valor award?" The rest of the President's guards, who are weaponless, ironically by presidential executive order, are quickly zip tied up and taken away. Jeff signals to one of his men to cover the mangled dead body on the ground with a tablecloth from one of the nearby tables.

Just as two of Jeff's men spread and lay down a white tablecloth over the body, a loud unidentifiable, unearthly tone begins to reverberate throughout all of Callisto. The soundwaves create a small earthquake that can be felt in every home. The large decorative windows all around the hall shatter into pieces, sending razor-sharp glass shards down into the room. The crystal chandeliers swing wildly from side to side. Everyone, including Jeff and his men, stops dead in their tracks and starts to panic, barely able to stand. You can hear the gasps from the crowd that is trying desperately to understand what is happening. They all cower in fear and cover their heads and eyes from the glass and falling debris. Jeff and his crew hold on to their rifles but are confused at what is going on. Even these hardened murderers are genuinely afraid and brought to their knees. The noise and the shaking suddenly stop. "What was that?" the President fearfully asks. "I don't know," Thomas says. Chairman Roth starts to quietly move toward the exit. "Where are you going?" Jeff asks. Mr. Roth puts his hands in the air and stops where he stands. Out of nowhere, a gigantic hand comes bursting through a hole in time and space through the main wall behind them. Jeff and the others all shift their eyes back to the front of the room above the head table where the President and his guests sit. The Adimowan Fallen all around Callisto immediately fall to the ground and start to moan and wail as they rock back and forth, sensing that this is judgment from the Highest. Their halos embedded in the back of their heads begin to brighten to the maximum brightness.

Everyone in the banquet hall turns to look at the marvelous wonder of a hand coming out of nowhere from a rip or tear into this dimension. The hand begins to write words on the wall in writing that every person, even from different nations can somehow read and understand. The President reads it to himself, "You have been tried in the balance and found wanting, your kingdom is divided." The same thing that the Highest wrote on the wall to King Belshazzar of the Kingdom of Babylon, in the Bible so long ago. The President turns to Gladys and asks, "What does that mean?" Gladys stands up from her seat with eyes widened and begins to rip her clothes with both hands. "I told you Randolph, hell to pay, hell to pay, hell to pay for us all!" Gladys shrieks. They all start to panic and scream, knowing that this is a spiritual judgment on the wall. Some think that it is all a part of some cleverly produced show and just sit there, wondering what will happen next to astonish them. Most of the crowd runs outside in terror and starts to look up, anticipating their doom.

Almost every person on Callisto hears the sound and stops whatever they are doing. Young and old that hear the tone are compelled to react to it. For some reason, children under the age of thirteen are not able to hear the sound. At that moment, the law of gravity on Callisto is suspended, just enough for humans and Adimowan fallen to begin to float away at a high rate of speed. Most struggle to hold onto something to keep them down, but the pull becomes too strong to fight Thousands are all snatched up and away, one by one into the night sky. People inside of buildings and houses rise and hit the ceilings. The force of reverse gravity eventually pulls them through the roof. It's as if they're being pulled up by strong magnetic forces.

Gladys just stares at Randolph as she is taken upward. She barely even fights the forces pulling her, knowing that this is God punishing and hell to pay as she said. The President holds on as long as he can just as his children run into the room and see him just as he lets go. His children Sophia, Mark, and his eldest son Gregory, all under thirteen, look up as their father floats up and away. Gregory stretches his hand up towards his father, reaching out to him. "Come back," Gregory says in a soft voice. In complete shock, they wave a solemn goodbye to him. With tears in his eyes and screaming silently into the darkness, The President of the United States of Callisto and his chosen are no more.

At that moment the small ball of bright light flows out of the President's dead body and comes to rest over Gregory's head. Sophia and Mark are gone and in complete shock, as they watch all of this happen before their eyes. Then the ball forces itself into Gregory's mouth and he is made to swallow it whole, as though the covenant with Pherous continues. Outside, Jeff Aragon and his team try to strap in and lock themselves down with cables and paracord. They scream as they are torn and snatched upward. Their toughness and pride are about to become a memory. "Momma!", Jeff screams as he looks back at the ground that is getting further and further away from him.

The skies are filled with the rich, wealthy, old money, haves that couldn't purchase their way out of this one. It looks something like a Chinese lantern festival gone wrong. They all just float up into the outer darkness of space. They begin to freeze in the cold as they go higher and higher. The ever-reaching judgment of the Highest has found them, proving that there is nowhere in any universe to hide. Tragically, some that float away elect to hold their children in a cold embrace as they leave Callisto. It's a terrifying scene out of a horror movie. In one hour, Callisto is quiet again, the trees rejoice. Only the animals, plants, and children of Callisto that were left behind remain. After everything goes quiet and quite surreal, the children wander outside begin to cry. The twelve-year-olds are in charge of everything now and instinctually move to take care of the little ones. They try to console them as they watch the adults, Men, women, and Adimowan alike, fly away.

CHAPTER 23

BACK TO THE RIVER

SECOND ADIMOWA, ETERNITY (NO TIME ZONE), CONTINUOUSLY DAY

Horusal and Makia smile at each other as the sunlight shines on their faces. Birhan, Fatimah, and the girls are stunned. They all stand and look all around in shock, wondering how this happened. Horusal remembers a scripture he once read in the bible, thinking this has to be its fulfillment. Revelations 21 "And I saw a new Heaven and a New Earth: For the first Heaven and the first Earth were passed away;" The emotions that they all feel at this moment cannot be described, words fail, true meaning hard to grasp. They finally realize that they are now living in the kind of story that the Bible records. The great battle has taken twenty-two hours to be fought.

There is still time to get back into the River of Life and go to Earth. "We need to get back there and try to help everyone else," Angelia says. "There is still time." "The world as we knew it is gone baby," Louis says. "We must try; I agree with Angelia," Zoe says. "What do you think Alfonso?" Zoe asks. "I want to stay here; I mean this is the goal, right?" "Birhan, Fatimah, what do you think?" asks Zoe. "I don't know," Birhan says. "What do you think Fatimah?" he asks. "We should go to Callisto." "Adim said that he would allow us to start over on Callisto, that is his will," Horusal says. He looks at Makia. "We will stay together my love." Rodrigal looks at them all as they decide what to do next. "His choice has already been made.

Makia, I and Aita will go to Callisto and start over as Adim told us. I would understand if you wanted to go back, but I'm not sure what you will be going back to. The Earth is lost."

"We'll go with you dad," says Louis. The kids all smile as they watch their parents hug each other. They have finally forgotten all about whatever it was that made them fall out of love. "We want to go too, with you to Callisto, the new Earth," they all say. "We are with you too," says Birhan. Zoe and Alfonso talk it over and decide to go back. "I want to try. I want to tell people, whoever is left, what I know. A different kind of reporting, I guess," Zoe says. "There's another reason too." She hesitates. "My Father died of COVID a few weeks ago, right before all this happened." Alfonso looks surprised to hear her say that. "You didn't even tell me that Zoe," he says. "I know. He was 83 and we hadn't spoken to each other in a while. Then he died. I didn't even go to the hospital or the funeral." They all look at her. "How could I do that? I hurt my mother bad with that one. I must go back and get things right somehow if they are even still alive." "I want to see about my family too," Alfonso says. "I have to know if they made it or not. I also want to be with you Zoe," he says, taking her by the hand. She allows him to take it. While looking deep into his eyes, she simply says, "Ok, Alfonso." "You better hurry," says Makia. "She's right," Horusal says. With no time to waste, everyone runs to the cable-less antigravity elevators with Alfonso and Zoe and goes back down to First Adimowa. They finally arrive at the banks of the River of Life.

A large crowd stands nearby to watch them go. They both look back at everyone standing around the river and then jump in with only minutes to spare. They look like they "jumped the broom" like at a traditional West African wedding. Their journey home to Earth begins. Zoe and Alfonso will return with a clear understanding of what is the truth and what is not. "How do we get to Callisto?" Josh asks Aita. "The Highest has already written it," Aita says, pointing to another River that flows down from the throne in Third Adimowa. "I will go with you," Aita says. Angelia and Makia hug her. "Grandma, I'm so happy that you're going with us," Angelia says. "I still have a lot of questions." Aita is overtaken with joy. She has never been called grandma before. Her halo begins to shine very brightly. "From now on Callisto and Adimowa will be the same, just as Earth once was," Aita says. "What do we need to do?" Horusal asks. The answer is there.

They all walk over together and look down into the crystal-clear river. Rodrigal stays behind knowing that he will not be allowed to go. They can see nothing at first; then they begin to see Callisto from space in the reflection on the water. The view quickly changes to the atmosphere above it, then the view on the ground. The terrain is lush, green, and perfect as the Garden of Eden. Oceans filled to the brim with life and skies with birds of every kind. Horusal smiles and points down in absolute wonder and complete amazement. "There it is. I can see it," Horusal says with fascination, looking over to Rodrigal who now stands on the other side of the stream. At that very moment, a deep, rich voice from behind calls out to them. "Let there be," the voice says. They all turn to the sound of the voice and find themselves instantly gone from Adimowa and standing in the middle of beautiful Callisto. "How did this happen?" Mona asks. "The Highest," Josh says, "it was him." Horusal falls to his knees in utter amazement and joy. He lives again and now with his true love and family all around.

CHAPTER 24

THE SEED OF CAIN

CALLISTO, SECOND LARGEST MOON OF JUPITAR, AUGUST 5, 2021

They all start to walk around and look up into the beautiful blue sky of Callisto. The kids are all the ages that they were when they translated from the Great Pyramid, but Horusal is still a healthy thirty-three just like Makia who stands beside him now shorter than him. They all lift their hands in grateful adoration to the Highest. Birhan and Fatimah run around and play like children. Aita is regular human size now, and stands about 5'9". She walks around and just takes everything in as she thinks about Makal and the changes to her body. The children of the lost chosen that remain start to come out of their hiding places and begin to surround them. "Who are they?" Sammy? asks. "They must be the children left behind," Makia says, "they were innocent."

Some of them stay off to themselves, still afraid of the new arrivals. Most are so glad to see adults again. Angelia and Louis embrace one another. Mona, Sammy, Josh, Gabriel, and Dellina go over to the other children and try to make them feel more comfortable. "We made it," Horusal says, as he turns to his love. "Yes, we did," Makia says. "What now?" Angelia asks. Horusal pauses for a few moments and looks up as if waiting for instructions. "We rebuild and replenish the new Earth. We do what He tells us to do and we live. His laws still stand. I will need everyone's help, especially you Birhan, and Louis." "We got you," Louis says, balling his fist and pounding his chest. "This is so wonderful, I will spend eternity helping you my brother, as long as you need it. I will be here for you," Birhan says "Thank you, old friend," Horusal says, "thank you all." Horusal and Makia embrace each other as they think of all the many years they were separated from each other. They look at everyone standing beside them and just smile and laugh. "I never thought that we would have this," Makia says, "the Highest is truly merciful."

"Let's find a place to live," Horusal says. "We have some catching up to do." Makia smiles at Horusal in that way. All of the many houses and condominiums are completely furnished and recently vacated and empty. "Let's live by the sea," Horusal says, I've always wanted to live there. "We love the sea too dad," Angelia says, "but only if you agree honey." "I love the sea and I love you," Louis says. Mona, Sammy, and Josh's hearts warm at the site of their parents not fighting anymore and even back in love again. They can barely remember the way it used to be back on Earth. "Father!" Makia yells as she sees Makal standing not too far in the distance, waving at them. The Highest has allowed him to live again with his complete family. He is a regular human size too. Aita sees him just as they all do. They cannot believe what they are seeing, looking at him like they've seen a ghost. "Lazarus has returned," Horusal says. They run to him as he runs to them with all his might. Makal cries out with tears of joy and gladness as he reaches his wife Aita and the rest of the family. Aita doesn't hesitate to fall on his neck and into his strong arms. Makia and the others grab him in a bear hug to end all bear hugs. They all collapse to the ground.

"I'm so glad that you are here," Horusal says in tears. "Now we are complete," Angelia says. They experience love and tenderness that surpasses all natural understanding. It is mingled with pure thankfulness and gratitude to the Highest for performing such wondrous miracles. The children that were left behind are happy too, hugging and crying. Over the next few months, they each distribute the children left behind equally into each family. Birhan and Fatimah only get two children, Bernard and Alexus. Angelia and Louis get four more children, Charles, James, Danny, and Linda. Aita and Makal have five, Rachel, Carlos, Cristina, Max and Juda. Horusal and Makia have taken on seven beautiful children. Salvador, Lizzy, Oliva, Joel, Mark, Gregory, and Sophia. Almost all of the children eventually adjust fairly well. Some struggle to deal with the loss of their parents and guardians and some never really get right, in spite of being in total paradise and having all the love a child could ask for.

Callisto

The Fourth Moon of Jupiter,

August 5, 2022, 4:29 PM (Space-Time)

Horusal, with a full white beard, sits down in a chair while Makia stands beside him. "It's been exactly a year today, Makia. Look at them," he says, looking down at their adopted children playing on the beach. "We are truly blessed to be here. He has given us all this," he says, pointing to the Western horizon. They both stand on the balcony of their modest home with an absolutely beautiful view of the ocean. Horusal puts his arm around Makia's shoulder and pulls her close. "Yes, we are very blessed," Makia says, "we are flourishing here as we were commanded. I know that it pleases the Highest to see that. The little ones are happy, they are adjusting well." "I believe they are," Horusal says. "All of the family is doing well. What's on your mind, my love? Do you miss Adimowa and the way you were, all tall and powerful?" he asks her. "Yes, sometimes but I would much rather be here with you small and just as powerful.

Everything that I love is here," Makia says with confidence. "But one day we will return to the River, Adimowa is home for us all. There we will live together forever with all our ancestors and dwell in the light of the Highest. The Highest brought you back and changed me, but we still must die one day." "I'm good with that," Horusal says. "Do you miss Earth?" she asks, "being who you were." "Wow, I guess I do a little bit but this place is so wonderful and different but the same. They are all very much the same because the Highest made them all." "That is true," Makia says. "I've been thinking, we must make sure that the children know how all this happened and what sacrifices were made. We will never forget or break the laws of the Highest. We cannot repeat what we did on Earth. Do you understand my dear?" Horusal asks. "Yes," Makia says. "You are wise to feel this way, my love. You teach them," she says.

From that day, Horusal begins to write down everything that happened in a series of leather-bound books. Over time he has volumes of books on his shelf. He has recorded the epic adventures of his life and times. He writes of how he first met Makia, how they had a baby named Angelia, how the family translated to Adimow through the Great Pyramids, and how they fought and won against Pherous and his wicked Father Lucipherous with the awesome help of the Adim. How they were allowed to start over in peace of Callisto. Horusal and Makia walk together into the town square and begin to gather the people and all who will listen to him. Most people in the town are wondering what he has to say to them that is so important. Makia beckons to them to come and sit around them.

After a few minutes, several hundred men, women, and children gather around. Someone gets the idea to ring the town's emergency bell to make sure everyone knows that there is something special happening. "Everyone, I want to tell you a true story," Horusal says. "An unbelievable story beyond all imagination." They all sit down and wait for Horusal to tell them the story. Horusal, now known as The Great Father of Callisto, stands in the middle slightly higher than the people and opens up the first book. The crowd goes reverently quiet as he stands there ready to begin. All eyes are on him as they wait for him to begin. It's sort of quickly turned into a festive-like atmosphere. He starts to read from the first book. Everyone is listening, learning, and happy or so it seems. One of Horusal and Makia's children stands off in the distance, just glaring so-called "The Great Father" with eyes full of hatred and malice. The kind of eyes that have vengeance in mind. He quietly rebels against his foster parents and has decided that he can't do this anymore. He remembers who his real father is. The seed of Cain, the unquenchable desire to kill your own brother, seems to live on in twelve-year-old Gregory, the eldest son of the former President of the United States of America.

The End.

CPSIA information can be obtained
at www.ICGtesting.com
Printed in the USA
BVHW051756010322
630321BV00012B/830